ANIMAL TRACKS of Texas

by Chris Stall

THE MOUNTAINEERS/SEATTLE

To Julia

The Mountaineers: Organized 1906 ". . . to explore, study, preserve, and enjoy the natural beauty of the outdoors."

4 3 2 1
5 4 3 2 1

Published by The Mountaineers
306 Second Avenue West, Seattle, Washington 98119

Published simultaneously in Canada by Douglas & McIntyre, Ltd.,
1615 Venables Street, Vancouver, B.C. V5L 2H1

Manufactured in the United States of America

Book design by Elizabeth Watson
Track on cover: Fox Squirrel

Library of Congress Cataloging in Publication Data

Stall, Chris.
Animal tracks of Texas / by Chris Stall.
p. cm.
Includes bibliographical references and index.
ISBN 0-89886-228-0
1. Animal tracks--Texas--Identification. I. Title.
QL768.S7328 1990 90-22136
591.9764--dc20 CIP

Contents

Preface

Most people don't get a chance to observe animals in the wild, with the exceptions of road kills and a few nearly tame species in parks and campgrounds. Many wild animals are nocturnal or scarce, and many are shy and secretive in order to avoid the attention of predators, or stealthy as they stalk their next meal. In addition, most wild creatures are extremely wary of humans either instinctively or because they've learned through experience to be that way. We may catch fortuitous glimpses now and then, but few of us have the time or motivation required for lengthy journeys into wild country for the sole purpose of locating animals. The result is that areas where we would expect to see animals often seem practically devoid of wildlife.

That's rarely the case, of course. Actually, many animals reside in or pass through all reasonably wild habitats. Though we may not see them, they nevertheless leave indications of their passage. But for the most part such signs are so obscure or confusing that only the most experienced and knowledgeable wilderness travelers notice them.

There's one grand exception: *animal tracks.* Often readily apparent even to the most casual and inexperienced observer, tracks not only indicate the presence of wild animals but can also be matched relatively easily with the animals that made them. I guess that's why I have been fascinated by animal tracks since my childhood in rural New York, and why that focus has continued through two decades of wandering and searching for them in wildlands all across North America.

Animal Tracks of Texas is a compilation of many years and many miles of my own field work, protracted observations, sketching, photography, and research into articles and books too numerous to list and too heavy to carry into the backcountry.

Animal tracks may be something you concern yourself with only when you happen on them, or your interest in tracks may become nearly obsessive. You may find yourself hiking with your chin resting securely on your chest, feverishly scanning the ground for clues. You may seek out snow because tracks show up on it better than most other surfaces. In the absence of snow, you might find yourself altering your routes, avoiding bedrock and ground cover, seeking out damp sand, soft dirt, and mud along streams, near ponds and lakes, around swamps. You may journey into the desert in the early morning, before the sand dries and moves on the wind. After a rainfall, you might make special trips to check fresh mud, even along dirt roads or hiking trails, knowing that among evidence of human activity the animal prints will be clear and precise.

Whatever your degree of interest, I hope you will enjoy using this book, in your backyard or in the wildest and most remote parts of Texas, and that your interest in identifying tracks grows until you reach the level of knowledge at which you no longer need this book.

Good luck!

Chris Stall
Cincinnati, Ohio

Introduction

HOW TO USE THIS BOOK

1. When you first locate unknown tracks, look around the immediate area to locate the clearest imprint (see Tracking Tips below). You can usually find at least one imprint or even a partial print distinct enough for counting toes, noting the shape of the heel pad, determining the presence or absence of claw marks, and so on.
2. Decide what kind of animal is most likely to have made the tracks, then turn to one of the two main sections of this book. The first and largest features mammals; the second, much shorter section is devoted entirely to birds.
3. Measure an individual track, using the ruler printed on the back cover of this book. Tracks of roughly 5 inches or less are illustrated life-size; those larger than 5 inches have been reduced as necessary to fit on the pages.
4. Flip quickly through the appropriate section until you find tracks that are about the same *size* as your mystery tracks. The tracks are arranged roughly by size from smallest to largest.

5. Search carefully for the tracks in the size range that, as closely as possible, match the *shape* of the unknown tracks.

6. If you find the right shape but the size depicted in the book is too big, remember that the illustrations represent tracks of an average *adult* animal. Perhaps your specimen was made by a young animal. Search some more; on the ground nearby you might locate the tracks of a parent, which will more closely match the size of the illustration.

7. Read the comments on range, habitat, and behavior, to help confirm the identification.

This book is intended to assist you in making field identifications of commonly encountered animal tracks. To keep the book compact, my remarks are limited to each animal's most obvious characteristics. By all means enhance your own knowledge of these track makers. Libraries and book stores are good places to begin learning more about wild animals. Visits to zoos with Texan wildlife on display can also be worthwhile educational experiences. And there's no substitute for firsthand field study. You've found tracks, now you know what animals to look for. Read my notes on diet, put some bait out, sit quietly downwind with binoculars for a few hours, and see what comes along. Or follow the tracks a while. Use your imagination and common sense, and you'll be amazed at how much you can learn, and how rewarding the experiences can be.

As you use this book, remember that track identification is an inexact science. The illustrations in this book represent average *adult* tracks on *ideal* surfaces. But many of the tracks you encounter in the wild will be those of smaller-than-average animals, particularly in late spring and early summer. There are also larger-than-average animals, and injured or deformed ones, and animals that act unpredictably. Some creatures walk sideways on occasion. Most vary their gait so that in a single set of tracks fore prints may fall ahead of, behind, or beneath the hind prints. In addition, ground conditions are usually less than ideal in the wild, and animals often dislodge debris, which may further confuse the picture. Use this book as a guide, but anticipate a lot of variations.

In attempting to identify tracks, remember that their size can vary greatly depending on the type of ground surface—sand that is loose or firm, wet or dry; a thin layer of mud over hard earth; deep soft mud; various lightly frozen surfaces; firm or loose dirt; dry or moist snow; a dusting of snow or frost over various surfaces; and so on. Note the surfaces from which the illustrations are taken and interpret what you find in nature accordingly.

You should also be aware that droplets from trees, windblown debris, and the like often leave a variety of marks on the ground that could be mistaken for animal tracks. While studying tracks, look around for and be aware of nonanimal factors that might have left "tracks" of their own.

The range notes pertain only to Texas. Many track makers in this book also live elsewhere in North America. Range and habitat remarks are general guidelines because both are subject to change, from variations in both animal and human populations, climatic factors, pollution levels, acts of God, and so forth.

The size, height, and weight listed for each animal are those for average adults. Size refers to length from nose to tip of tail; height, the distance from ground to shoulder.

A few well-known species have been left out of this book: moles and bats, for example, which generally leave no above-ground tracks. Animals that may be common elsewhere but are rare, or occur only in the margins of Texas, have also been omitted. Some species herein, particularly small rodents and birds, stand as representatives of groups of related species. In such cases the featured species is the one most commonly encountered and widely distributed. Related species, often with similar tracks, are listed in the notes. If their tracks can be distinguished, guidelines for doing so are provided.

If you encounter an injured animal or an apparently orphaned infant, you may be tempted to take it home and care for it. Do not do so. Instead, report the animal to local authorities, who are better able to care for it. In addition, federal and state laws often strictly control the handling of wild animals. This is always the case with species classified as *rare* or *endangered*. Removing animals from the wild may be illegal.

TRACKING TIPS

At times you'll be lucky enough to find a perfectly clear and precise track that gives you all the information you need to identify the maker with a quick glance through this book. More often the track will be imperfect or fragmented. Following the tracks may lead you to a more readily identifiable print. Or maybe you have the time and inclination to follow an animal whose identity you already know in order to learn more about its habits, characteristics, and behavior.

Here are some tips for improving your tracking skills:

1. If you don't see tracks, look for disturbances—leaves or twigs in unnatural positions, debris or stones that appear to have been moved or turned. Stones become bleached on top over time, so a stone with its darker side up or sideways has recently been dislodged.

2. Push small sticks into the ground to mark individual signs. These will help you keep your bearings and "map out" the animal's general direction of travel.

3. Check immovable objects like trees, logs, and boulders along the route of travel for scratches, scuff marks, or fragments of hair.

4. Look at the ground from different angles, from standing height, from kneeling height, and if possible, from an elevated position in a tree or on a boulder or rise.

5. On very firm surfaces, place your cheek on the ground and observe the surface, first through one eye, then the other, looking for unnatural depressions or disturbances.

6. Study the trail from as many different directions as possible. Trail signs may become obvious as the angle of light between them and your eyes changes, especially if dew, dust, or rain covers some parts of the ground surface.

7. Check for tracks beneath recently disturbed leaves or fallen debris.

8. Try not to focus your attention so narrowly that you lose sight of the larger patterns of the country around you.

9. Keep your bearings. Some animals circle back if they become aware of being followed. If you find yourself following signs in a circular path, try waiting motionless and silent for a while, observing behind you.

10. Look ahead as far as possible as you follow signs. Animals take the paths of least resistance, so look for trails or runways. You may even catch sight of your quarry.

11. Animals are habitual in their movements between burrows, den sites, sources of water and food, temporary shelters, prominent trees, and so on. As you track and look ahead, try to anticipate where the creature might be going.

12. Stalk as you track; move as carefully and quietly as possible.

The secrets to successful tracking are patience and knowledge. Whenever you see an animal leaving tracks, go look at them and note the activity you observed. When you find and identify tracks, make little sketches alongside the book's illustrations, showing cluster patterns, or individual impressions that are different from those drawn. Make notes about what you learn in the wilds and from other readings. Eventually, you will build a body of knowledge from your own experience, and your future attempts at track identification will become easier and more certain.

This book is largely a compilation of the author's personal experiences. Your experiences with certain animals and their tracks may be identical, similar, or quite different. If you notice a discrepancy or find tracks that are not included in this book, carefully note your observations, or even amend the illustrations or text to reflect your own experiences.

Mammals
Reptiles
Amphibians
Invertebrates

INVERTEBRATES

The smallest track impressions you are likely to encounter in nature will probably look something like those illustrated at the right.

From left to right, the illustration shows tracks of two common beetles, a centipede, and a cricket. The track of an earthworm crosses from lower left to the upper right corner.

You might initially mistake a variety of scuffs and scratches left by windblown or otherwise dislodged debris, the imprint of raindrops that have fallen from overhanging limbs, impressions left by the smallest mice, or even the perplexing calligraphy of toads for insect marks. If you have more than a square foot or so of ground surface to scrutinize, however, you will usually find that insect tracks form a recognizably connected line; the extremely shallow depth of the trail of imprints is also a good clue that a very lightweight being has passed by.

With literally millions of species out there, trying to identify the insect that made a particular track can be challenging, but there are times when you can follow a trail and find, at the end, either the bug itself, or a burrow that could yield its resident with a little patient and careful excavation on your part. If you spend enough time in one area, you will begin to observe specific species in the act of making their tracks, and that goes a long way toward track recognition—for any size of animal.

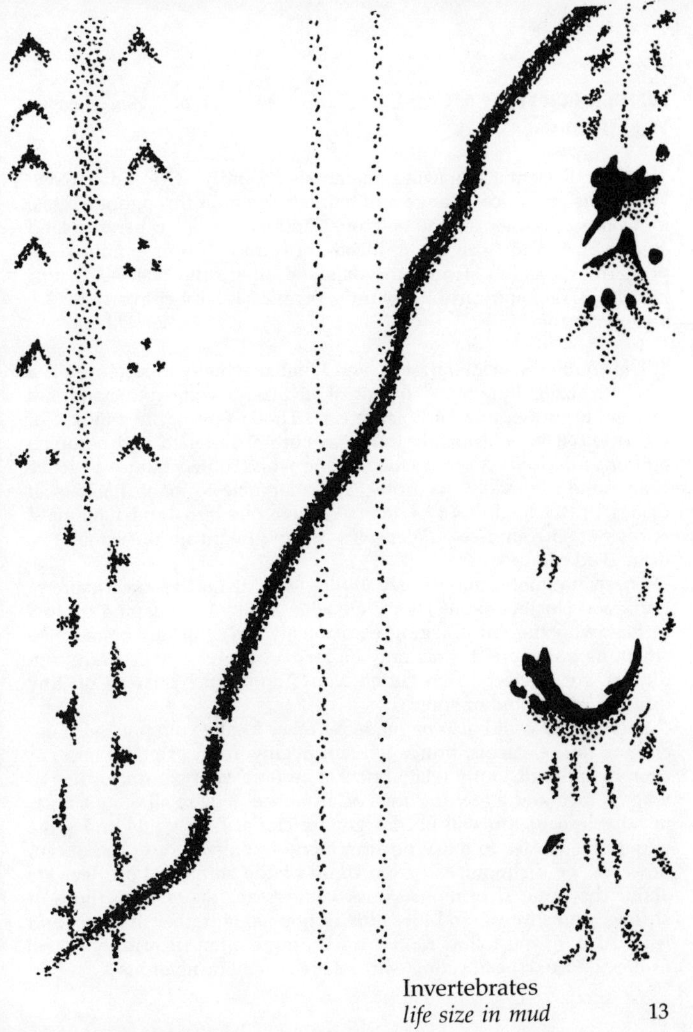

Invertebrates
life size in mud

WHITE-FOOTED MOUSE *Peromyscus leucopus*
Woods mouse

Order: Rodentia (gnawing mammals). **Family:** Cricetidae (New World rats and mice). **Range and habitat:** common throughout Texas; in woods, prairies, rocky outcrops, and nearly all other dry-land areas. **Size and weight:** 8 inches; 1 ounce. **Diet:** omnivorous, primarily seeds but also mushrooms and other fungi, berries, herbs, insects, larvae, and carrion. **Sounds:** occasional faint chirps, squeaks, and chattering.

The abundant, wide-ranging, and familiar white-footed mouse is a medium-sized, long-tailed mouse with pretty white underside and feet, pointy nose, and fairly large ears. Though primarily a mouse of the wilds, it is occasionally found in both abandoned and occupied buildings as well. A good climber, the white-footed mouse is active year-round, generally nocturnal, and adaptable to many habitats. It makes up the main diet of many carnivorous birds and mammals, but is not so completely defenseless as you might think—it may bite if handled carelessly.

The white-footed mouse usually leaves a distinctive track pattern: four-print clusters about 1½ inches wide, walking or leaping up to 9 inches, with the tail dragging occasionally. As with any creature so small, its tracks are distinct only on rare occasions when surface conditions are perfect. More often you find vague clusters of tiny dimples in the mud or snow.

Such tracks could also be made by other locally common mice, including house, cactus, brush, golden, pygmy, rock, or piñon mice, or even by more distantly related pocket and grasshopper mice. Give or take an inch and a few fractions of an ounce, they're all quite similar in appearance. You will need a good pictorial field guide and some patient field work to make positive identifications. Mouse tracks can, however, be distinguished from those of the shrew, whose feet are about the same size: mouse track clusters are wider than those of shrews, and shrews tend to scuttle or hop along, rather than run and leap. Also, if you follow mouse tracks, more often than not you will find evidence of seed eating; shrews are strictly carnivorous.

White-footed Mouse
life size in mud

DESERT SHREW *Notiosorex crawfordi*
Gray shrew

Order: Insectivora (insect-eating mammals, including shrews, moles, and bats). **Family:** Soricidae (shrews). **Range and habitat:** throughout western and central Texas; in arid country, especially in chaparral and low-desert shrubs. **Size and weight:** 5 inches; 1 ounce. **Diet:** slugs, snails, spiders, insects, and larvae; occasionally mice and carrion. **Sounds:** commonly silent.

The desert shrew is a vole-shaped creature with shorter legs, a slightly more elongated body, and a long, pointed snout. Shrew dentists have a big advantage in distinguishing the various species, because variations in unicuspid teeth are all that differentiate many of them. *All* shrews are little eating machines, though, with extremely high metabolisms; in fact, shrews consume more than their own body weight in food on a daily basis.

Shrews' constant and aggressive quest for food makes their tracks, in general, fairly easy to identify. The animals move around with more single-minded purpose than mice or voles, usually in a series of short hops in which the hind feet fall over the tracks of the forefeet and the tails often drag, leaving the distinctive pattern shown, usually less than an inch in width. When individual impressions are more distinct, you may notice that shrews have five toes on both fore and hind feet (most mouselike creatures have four toes on the forefeet).

Similar tracks in moist habitats or grassy areas of eastern Texas may have been left by either least or southern short-tailed shrews. Wherever you find them, you'll see that the generic shrew track pattern is recognizable.

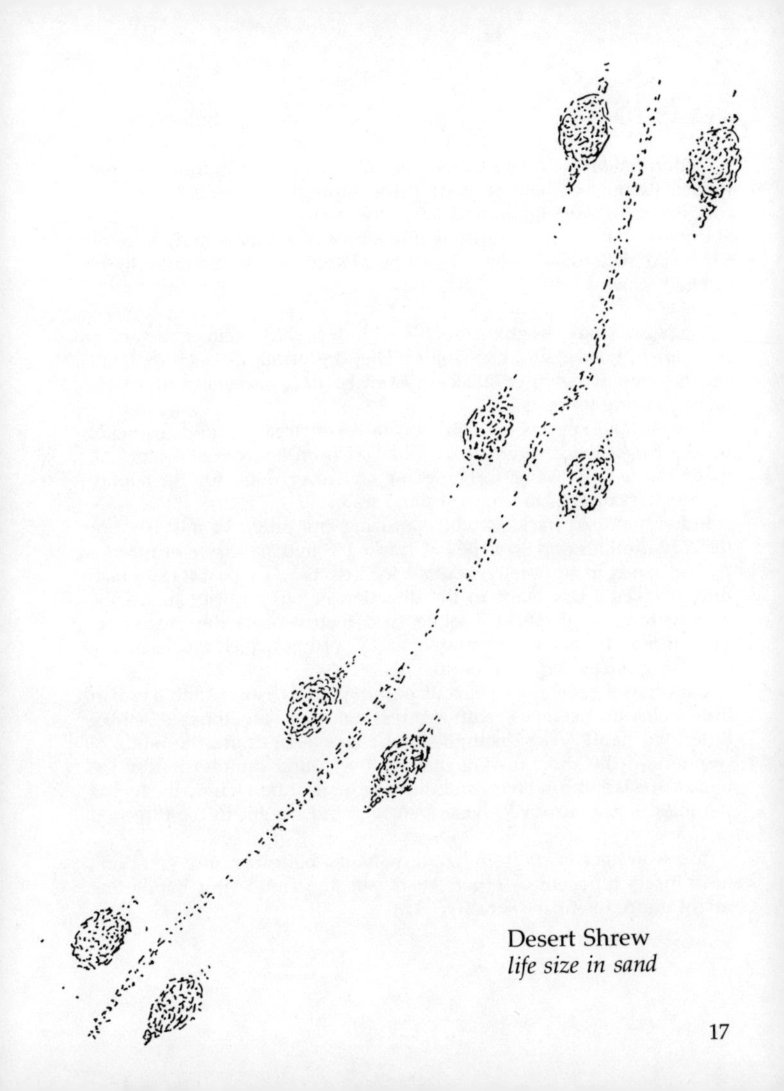

Desert Shrew
life size in sand

TEXAS TOAD *Bufo speciosus*

Order: Salientia (frogs, toads, and allies). **Family:** Bufonidae (true toads). **Range and habitat:** widespread throughout western and central Texas; in woodlands and adjoining areas, wherever insects are abundant, but not necessarily within a mile of permanent dampness. **Size and weight:** 4 inches; 2 ounces. **Diet:** insects. **Sounds:** high-pitched musical trills.

Toads are small, froglike animals with dry, warty skin, in a variety of reddish, brown, and gray colors. They are primarily nocturnal, but can be seen at dawn or dusk, or even by day, crouching in a little niche, waiting for bugs.

Several toad species are common in many areas around the state. Unlike frogs, toads often travel fairly far from sources of water. All toads do require water for breeding, however; look for their long, ropy strings of eggs in stagnant pond water.

Individual toad tracks can be confusing and might be mistaken for the tiny dimples and scratches of tracks left by small mice or insects. A toad tends to sit quietly, waiting for insects to fly past it, at which time it takes a few leaps in the direction of wing noise, snares the bug with its long, sticky tongue, and then repeats the procedure. Thus it may change direction over earlier prints, which makes a very confusing picture on the ground.

Toad tracks generally consist of nothing more distinct than a trail of little holes and scrapes, with impressions that sometimes resemble little toad hands. The distinguishing characteristics are the mode of wandering, the short rows of four or five round dimples left by the toes of the larger hind feet, and the drag marks often left by the feet as the toad moves forward; those toe drag marks point in the direction of travel.

You won't get warts from handling toads, but make sure you don't have insect repellent or other caustic substances on your hands that might injure the toad's sensitive skin.

Texas Toad
life size in mud

HISPID COTTON RAT *Sigmodon hispidus*

Order: Rodentia (gnawing mammals). **Family:** Cricetidae (New World rats and mice). **Range and habitat:** throughout Texas; in grassy and weedy areas. **Size and weight:** 14 inches; 2 ounces. **Diet:** insects, fungi, fruits. **Sounds:** silent.

The hispid cotton rat is a medium-sized, grayish brown, very abundant rat that often causes extensive damage to sugar cane and sweet potato crops and occasionally eats the eggs and chicks of quail. It is a generally nocturnal and solitary little beast that you probably will not sight very often.

Tracks similar to those illustrated but located in damp or marshy habitats of eastern Texas may have been left by a marsh rice rat, which is about the same size. As a general rule, however, within Texas any given habitat might be home to a great variety of similar little mouselike creatures, and the cunning use of a live trap may be the only practical way to determine which species of various young or full-grown voles, mice, or rats is actually leaving tracks in a particular area. Print lengths of up to ¾ inch and walking strides of about 2 inches may help to distinguish the larger rats.

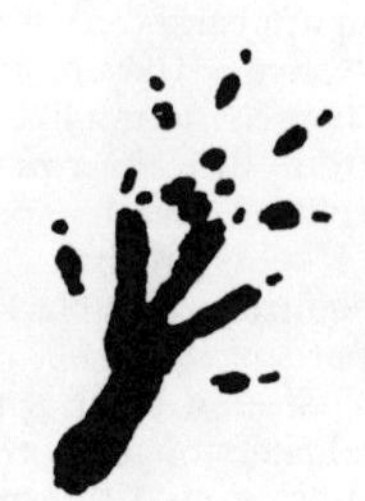

Hispid Cotton Rat
life size in mud

THIRTEEN-LINED GROUND SQUIRREL Striped gopher

Spermophilus tridecemlineatus

Order: Rodentia (gnawing mammals). **Family:** Sciuridae (squirrels). **Range and habitat:** throughout eastern Texas; in shortgrass prairies, pastures, lawns, cemeteries, and golf courses. **Size and weight:** 9 inches; 8 ounces. **Diet:** omnivorous, including herbs, seeds, fruits, insects, eggs, carrion. **Sounds:** variety of shrill and rapid chirps and birdlike whistles to signal alarm.

The five species of ground squirrels inhabiting Texas vary greatly in appearance, but they are all terrestrial, living in extensive burrow systems. They tend to stand up and whistle when alarmed, rather than chattering like their arboreal cousins, and they never venture far from their burrow entrances. Thirteen-lined ground squirrels are solitary, slender, pale-gray animals with prominent stripes or rows of spots on their sides. Mexican and spotted ground squirrels, rock squirrels, and black-tailed prairie dogs are also common in various parts of the state.

Ground squirrel tracks are always found near excavated burrows. Clusters of four prints separated by 9 to 18 inches are made by running ground squirrels—these frenetic little animals rarely walk. Another key feature of ground squirrel tracks is the toe prints. Ground squirrel toes, adapted for digging, tend to leave splayed prints. Their claws are straighter and longer than those of tree squirrels, and often leave imprints farther from the toes. One last difference: ground squirrels almost never leave tracks in snow; they are too busy sleeping away the winter.

Thirteen-lined Ground Squirrel
life size in mud

SOUTHERN PLAINS WOODRAT *Neotoma micropus*
Pack rat, trade rat

Order: Rodentia (gnawing mammals). **Family:** Cricetidae (New World rats and mice). **Range and habitat:** widespread throughout western and southern Texas; in semiarid brushland, thickets, and cactus areas. **Size and weight:** 15 inches; 1¼ pounds. **Diet:** prickly pear leaves, cacti, seeds, nuts, sotol, and agave vegetation. **Sounds:** silent.

This native American rat is active year-round but it is seldom sighted in the wild, as it is generally nocturnal. Woodrats are slightly larger than the imported Norway and black rats, and have furry, not scaly, tails. These rodents avoid human habitations, although they may "borrow" shiny objects from campsites.

The southern plains woodrat often constructs large houses, sometimes 4 to 5 feet high, near a cactus; or you may find piles of sticks, thorns, or other debris, with slender woodrat trails radiating from them. White-throated woodrats also live in parts of Texas, and eastern woodrats live in the hedges and wooded areas of eastern Texas. All are pale grayish brown, ratlike in appearance, and similar in size.

All woodrats have fairly stubby toes, four on the forefeet and five on the hind feet, which usually leave uniquely shaped tracks, slightly larger than cotton rat tracks, with no claw marks. The tracks are roughly in line when walking, and grouped as illustrated when running, with 8 or more inches separating the clusters of prints. Like most small rodents, when woodrats leap, their forefeet land first, followed by the hind feet, which come down ahead of the forefoot imprints, providing the spring into the next leap. If the characteristic stubby-toed imprints are not clear, the short spacing relative to foot size should help distinguish woodrat tracks from similar ones made by other animals of comparable size.

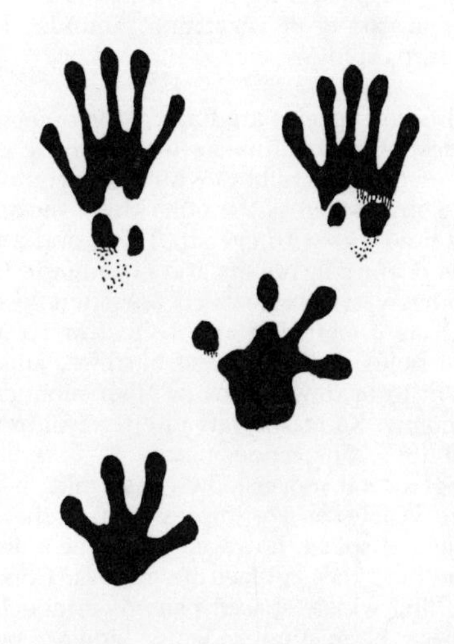

Southern Plains Woodrat
life size in mud

ORD'S KANGAROO RAT *Dipodomys ordii*

Order: Rodentia (gnawing mammals). **Family:** Heteromyidae (kangaroo mice and rats). **Range and habitat:** absent from central and eastern Texas; on sandy soil, dune areas, and hard-packed ground. **Size and weight:** 11 inches, much of which is tail; 2½ ounces. **Diet:** seeds and a great variety of vegetation. **Sounds:** thumps feet when alarmed, and chirps softly.

Three kangaroo rat species are found in various parts of the state. Ord's is a widespread, medium-sized rat, whose characteristics are representative of the lot: chubby, with buff or gray fur, white belly and feet, white stripes across the outer thigh meeting at the base of the tail, strong hind legs with oversized feet, and a very long, bushy-tipped tail. It is nocturnal, prefers arid or semiarid habitats, and lives in extensive burrow systems, which are often—especially in sandy areas—as much as 3 feet high and 12 feet or so in diameter, with many entrance holes. Around these burrows, and along the paths connecting them to feeding places or other mounds, you often find cut plant fragments. Kangaroo rats rarely travel more than about 50 feet from these not inconspicuous nests.

When a kangaroo rat moves slowly, as while feeding, most of its hind prints are visible, as are impressions of the smaller forepaws and the long tail. At speed, however, when the animal is leaping 3 to 9 feet, only the hind toes contact the ground. Combined with other signs, the resulting widely spaced pairs of prints clearly indicate the passage of these interesting rodents. Similar, but smaller, more closely spaced tracks are made by pocket and kangaroo mice.

Kangaroo rat tracks are easy to identify, once the basic pattern is discovered. Ord's is the only five-toed kangaroo rat in the state. Similar-sized four-toed prints in north-central Texas were left by a Texas kangaroo rat (*D. elator*), while larger four-toed prints in western Texas tell of the passage of a banner-tailed kangaroo rat (*D. spectabilis*).

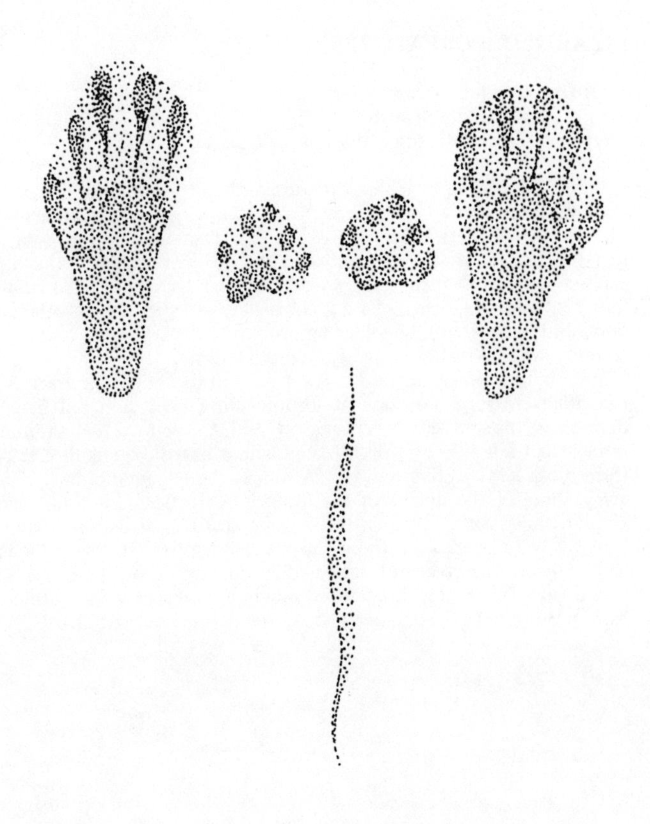

Ord's Kangaroo Rat
life size in mud

LIZARDLIKE CREATURES

Orders: Caudata (salamanders)
Squamata (lizards)
Crocodylia (crocodiles and alligators)

Many interesting lizardlike creatures inhabit portions of Texas, including the potentially dangerous American alligator, the largest reptile in North America, which lives in east Texas and can grow to 19 feet (an animal that could put a blight on any cookout); the tiger salamander, at 13 inches the world's largest terrestrial salamander, living throughout Texas; and a great variety of other newts, salamanders, and lizards, in a bewildering array of colors, shapes, and sizes, from less than 3 inches to nearly 2 feet in length.

You should, however, find it fairly easy to recognize the track of a lizardlike creature. Portions of its low-slung belly and tail usually drag along the surface it's walking on, and its five-toed feet alternate, rather than fall side by side, distinguishing lizardlike creature tracks from those left by other animals. At times, the waving tail may brush away some of the details of the footprints. Turtles sometimes leave similar trails, but a turtle with the same size feet as a lizard usually leaves a wider trail, with footprints closer together. The straddle and gait of lizardlike creatures vary widely depending on their size and speed, of course; my illustration shows a lizard of about 8 inches from nose to tail tip, such as the ubiquitous Texas horned lizard.

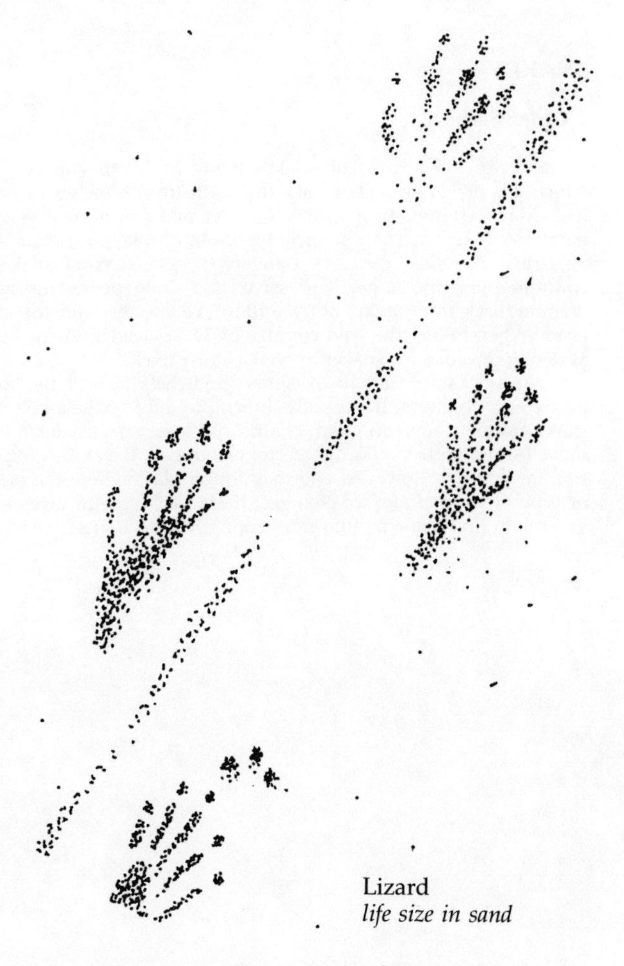

Lizard
life size in sand

SNAKES

Order: Squamata

There are 115 species of snakes living in North America, 19 of which are poisonous. They vary in length from 6 inches to nearly 9 feet. Many species, in a variety of sizes and colors, live in various habitats of Texas. The list includes most of the poisonous snakes of North America: the very dangerous eastern coral snakes and cottonmouths; the large, aggressive, and quite poisonous western diamondback rattlesnake; other rattlesnake species; and the copperhead. When hiking the wild country of Texas, watch where you step and look carefully for snakes as well as their tracks!

Snake tracks are easy to recognize, and the width of the snake is pretty easy to guess. It's usually difficult to tell a snake's direction of travel from its trail over flat ground, because snakes don't usually move fast enough to dislodge peripheral debris. If you can follow the trail far enough, however, you may find a place where the elevation or type of ground surface changes; there, with careful investigation, you might find some minute clues about the direction of travel.

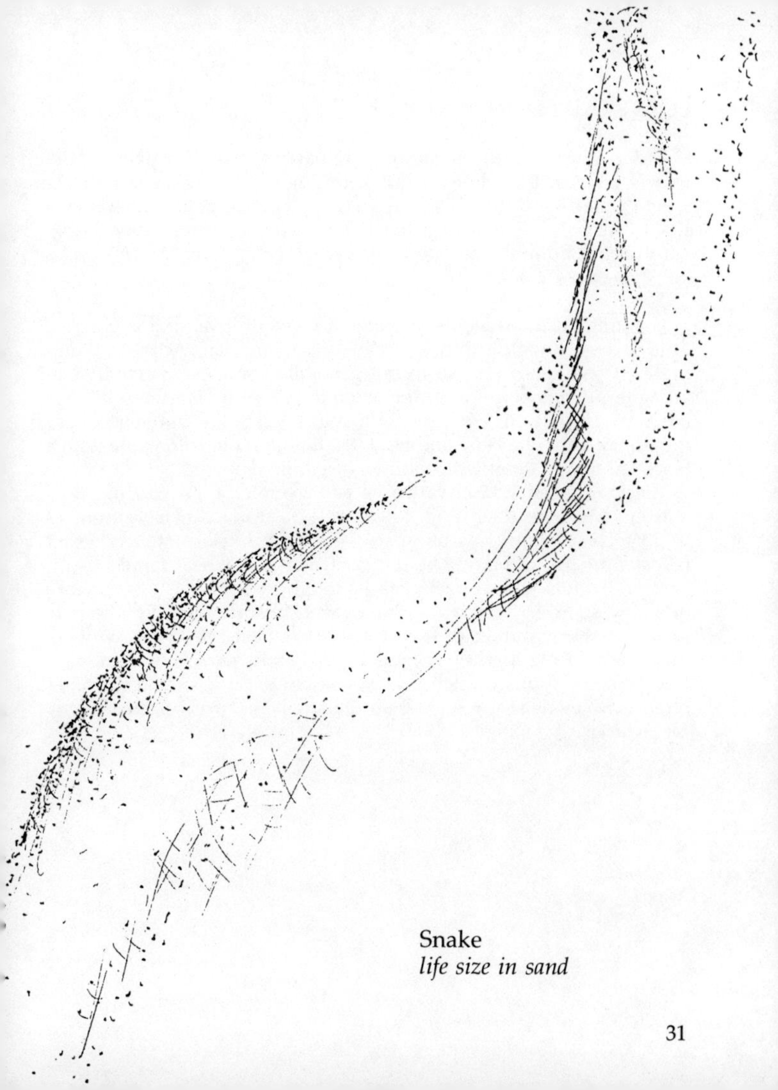

Snake
life size in sand

LONG-TAILED WEASEL *Mustela frenata*

Order: Carnivora (flesh-eating mammals). **Family:** Mustelidae (the weasel family). **Range and habitat:** throughout Texas; most habitats where water is nearby. **Size and weight:** variable, generally averages 12–18 inches; 6–9 ounces. **Diet:** small rodents, chipmunks, birds, amphibians. **Sounds:** may shriek or squeal when alarmed or making a kill; also purrs, chatters, hisses.

The long-tailed weasel is an adaptable, inquisitive, and aggressive little carnivore with a thin, elongated body and tail. Active day and night, it will climb and swim but generally confines its activities to an agile pursuit of prey on the ground, where it also finds various burrows. In portions of Texas with cold winters, its summer coat of brown with a white belly and black tail tip often molts to white with a black tail tip, or a mottled condition of partial molt.

The long-tailed weasel varies its gait frequently, alternating leaps with a variety of other modes of travel. Its stride is usually from 12 to 20 inches, with leaps of up to 50 inches. If close study doesn't reveal fifth toe prints, often the case with all weasel family members, individual weasel tracks can be tough to distinguish from those of many squirrels or rabbits. But weasels characteristically alternate long and short bounds, or leave lines of doubled-over tracks with occasional tail drag marks, whereas rabbits and squirrels tend to leave four separate prints in each cluster without tail drags. And the latter often leave evidence of vegetarian diets, whereas weasels are strictly carnivorous.

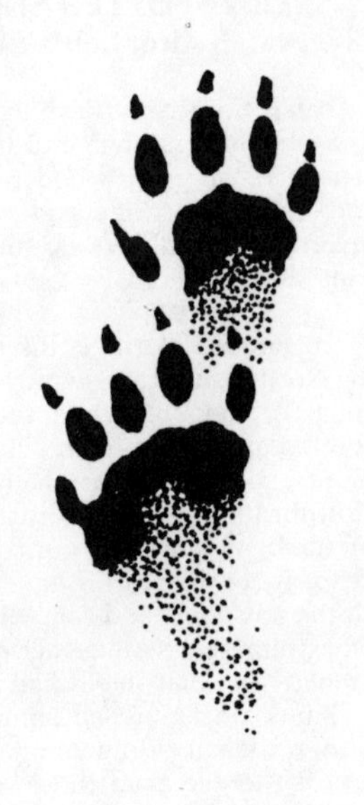

Long-tailed Weasel
life size in mud

EASTERN SPOTTED SKUNK *Spilogale putorius*
Civet cat, hydrophobia cat

Order: Carnivora (flesh-eating mammals). **Family:** Mustelidae (the weasel family). **Range and habitat:** widespread throughout Texas; in brushy or sparsely wooded areas along streams, among boulders, and in prairies. **Size and weight:** 25 inches; 2 pounds. **Diet:** omnivorous, including rats, mice, birds, insects, eggs, carrion, seeds, fruit, and occasionally vegetation. **Sounds:** usually silent.

The spotted skunk is the smallest and most visually interesting of the North American skunks. About the size of a small house cat, with an assortment of white spots and streaks over its black coat, the spotted skunk has finer, silkier fur than other skunks, is quicker and more agile, and occasionally climbs trees, although it doesn't stay aloft for long. It is primarily nocturnal, but you might see it at dawn or dusk, or foraging during the daylight in winter, when hunger keeps it active. Skunks have the most highly effective scent glands of all the mustelids and can, when severely provoked, shoot a fine spray of extremely irritating methyl mercaptan as far as 25 feet. Everyone knows what that smells like.

Skunk tracks are all similar, with five toes on each foot leaving prints, toenail prints commonly visible, and forefoot tracks slightly less flat-footed than those of the hind feet. Only size and irregular stride may help distinguish the tracks of the spotted skunk from those of its larger cousins. Spotted skunk tracks will be about 1¼ inches long at most; adult striped skunks leave tracks up to 2 inches in length. On the other hand, the quick spotted skunk leaves a foot or more between *clusters* of prints when running, while the larger species lope along with only about 5 or 6 inches between more strung-out track groups.

Because skunks can hold most land animals at bay with their formidable scent, owls are their chief predators. If you are following a skunk trail that ends suddenly, with perhaps a bit of black and white fur remaining mysteriously where the tracks disappear, you might be able to guess what transpired.

Eastern Spotted Skunk
life size in mud

SOUTHERN FLYING SQUIRREL *Glaucomys volans*

Order: Rodentia (gnawing mammals). **Family:** Sciuridae (squirrels). **Range and habitat:** portions of eastern and central Texas; in coniferous and, occasionally, mixed forests at higher altitudes. **Size and weight:** 11 inches; 6 ounces. **Diet:** bark, fungi, lichen, seeds, insects, eggs, and carrion. **Sounds:** generally silent; occasionally makes chirpy, birdlike noises.

Southern flying squirrels are nocturnal, so chances are that you will only see their tracks, unless you happen to knock against or cut down one of the hollow trees in which they are fond of nesting; in that case, if a small gray squirrel runs out, you've had a rare glimpse of the northern flying squirrel.

In warm areas, the southern flying squirrel doesn't leave much evidence of its passage. It lives mostly in trees, using the fur-covered membrane that extends along each side of its body from the forelegs to the hind legs to glide between trees and occasionally from tree to earth, where it usually leaves no marks on the ground cover of its forest habitat. If snow has fallen, however, you may find its tracks leading away from what looks like a miniature, scuffed snow-angel, the pattern left when the squirrel lands at the end of an aerial descent. The tracks may wander around a bit if the squirrel has foraged for morsels, but they will lead back to the trunk of a nearby tree before long.

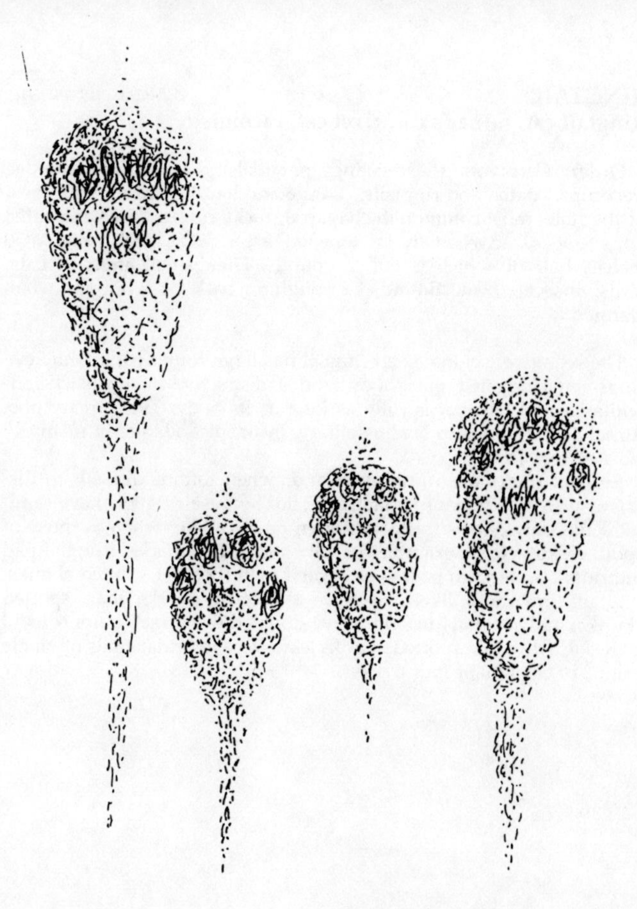

Southern Flying Squirrel
life size in sand

RINGTAIL *Bassariscus astutus*
Ringtail cat, miner's cat, civet cat, cacomistle

Order: Carnivora (flesh-eating mammals). **Family:** Procyonidae (raccoons, coatis, and ringtails). **Range and habitat:** throughout most of the state; most common in chaparral, rocky ridges, caves and cliffs, talus slopes, occasionally in forested areas. **Size and weight:** 30 inches, half of which is tail; 2 pounds. **Diet:** small animals, bats, birds, insects, fruit. **Sounds:** a coughing bark or whimper when alarmed.

The secretive, seldom seen ringtail has large round ears, white eye rings, an elongated gray body, and a distinctive bushy black-and-white-striped tail that is fully as long as its body. This strictly nocturnal animal tends to lead a solitary, inconspicuous life in relatively isolated terrain.

Ringtail tracks are uncommon and, when found, difficult to distinguish from others of similar size. But because ringtails have semi-retractile claws, their tracks are often quite catlike, whereas those of spotted skunks, for example, usually show claw marks. Ringtail pad imprints also tend to be less elongated than those of spotted skunks; ringtails are more likely to leave tail brush marks than spotted skunks; and ringtails usually leave doubled-over track pairs 6 to 10 inches apart, while spotted skunks leave very irregular trails of single prints, or clusters of four tracks.

Ringtail
life size in mud

MINK *Mustela vison*

Order: Carnivora (flesh-eating mammals). **Family:** Mustelidae (the weasel family). **Range and habitat:** most of eastern and central Texas; in brushy or open forested areas along streams, lakes, and other wetlands. **Size and weight:** 24 inches; 3 pounds. **Diet:** primarily muskrats and smaller mammals; also birds, frogs, fish, crayfish, and eggs. **Sounds:** snarls, squeals, and hisses.

The mink is the sleekest, most exuberant of the weasels, and the most aquatic. About the size of a small cat and medium brown all over, it is an excellent swimmer and may wander several miles a day searching for food along stream- and riverbanks and around the shorelines of lakes. Its den, too, is usually in a stream- or riverbank, an abandoned muskrat nest, or otherwise near water. Generally a nocturnal hunter, its tracks are likely to be the only indication you will have of its presence.

The mink leaves either groups of four tracks like those illustrated or the characteristic double pair of tracks, usually not more than 26 inches apart. The tracks nearly always run along the edge of water. Although the mink has five toes both fore and hind, it is quite common for only four-toed imprints to be apparent. Like all mustelids, the mink employs its scent glands to mark territory; so, as you track it through its hunting ranges, you may notice a strong scent here and there, different from but as potent as that of its relative the skunk. You might also find, in snow, signs of prey being dragged, invariably leading to the animal's den.

Tracks like these in the dirt of arid places or prairie-dog towns, however, are evidence of North America's rarest mammal, the endangered black-footed ferret (*M. nigripes*). This mink-sized mustelid is an active daylight hunter of prairie dogs and other rodent residents of its prairie habitat, and its prominent markings of dark mask, feet, and tail tip with tawny body and white face are easily identifiable. If you observe the area quietly from a distance, with binoculars if possible, you may be lucky enough to see one of these rare weasels.

Mink
life size in mud

STRIPED SKUNK *Mephitis mephitis*

Order: Carnivora (flesh-eating mammals). **Family:** Mustelidae (weasels and skunks). **Range and habitat:** throughout Texas; in semi-open country, mixed woods, brushland, and open fields. **Size and weight:** 24 inches; 10 pounds. **Diet:** omnivorous, including mice, eggs, insects, grubs, fruit, carrion. **Sounds:** usually silent.

This cat-sized skunk is easily recognized by the two broad stripes running the length of its back, meeting at head and shoulders to form a cap. A thin white stripe runs down its face. Active year-round, it is chiefly nocturnal. It is often found dead along highways, but live specimens may be sighted shortly after sunset or at dawn, snuffling around for food. It seeks shelter beneath buildings as well as in ground burrows or other protected den sites, and protects itself, when threatened, with a fine spray of extremely irritating methyl mercaptan.

Striped skunk tracks are similar to but larger than those of the smaller spotted skunk, and its elongated heel pads are often apparent because it's not as quick, agile, or high-strung as its smaller cousin. With five closely spaced toes and claws on all feet usually leaving marks, its tracks can't be mistaken for any other species of its size. The spacing is distinctive too: generally less than 6 inches between track groups, whether they consist of walking pairs or strung-out loping groups of four tracks. However, striped skunks in southern portions of the state share their territory with two very similar-sized cousins. Tracks like these along streams and among rocky ledges in southwestern Texas may have been left by a hooded skunk. The slightly larger hog-nosed skunk lives in partly wooded, brushy, or rocky parts of central and southern Texas. All of the larger skunks have similar modes of travel. Patient field observation is the only way to be certain which species is leaving tracks at any given site.

Striped Skunk
life size in mud

GRAY SQUIRREL *Sciurus carolinensis*

Order: Rodentia (gnawing mammals). **Family:** Sciuridae (squirrels). **Range and habitat:** eastern and central Texas; in hardwood (especially oak), pine, or mixed hardwood–evergreen forests and parklands, occasionally nearby in swamp fringes. **Size and weight:** 20 inches; 1½ pounds. **Diet:** mostly acorns and cone seeds, also various nuts, fungi, insects and larvae, some vegetation. **Sounds:** variety of rapid, raspy barks.

This large, light gray squirrel with its long upraised bushy tail is such a common park animal that most Texans are familiar with it. Active all day, year-round, the gray squirrel nests in tree cavities or in conspicuous nests made of sticks, usually 20 feet or more above the ground. It spends a lot of time on the ground searching for nuts and seeds, and ranges widely from its home trees.

Because of its wandering habits, the squirrel leaves a lot of tracks in areas traveled through by other similar-sized animals. Consequently, partial track impressions can be confusing. Most of the time, the squirrel scampers rather than walks, so its long hind heels don't leave prints. The real keys to recognizing the tracks of a gray squirrel are the number of toes—four on the forefeet, and five on the hind feet—and general track characteristics common to all members of the squirrel family. First, impressions of the entire toes, rather than just the tips, are often present. Second, the two middle toes of the forefeet and the three middle toes of the hind feet are nearly always out ahead of, and parallel to, the heel pads, with the outer toes splayed out to the sides. Gray squirrel track clusters are normally about 4½ inches wide and spaced 2 to 3 feet apart.

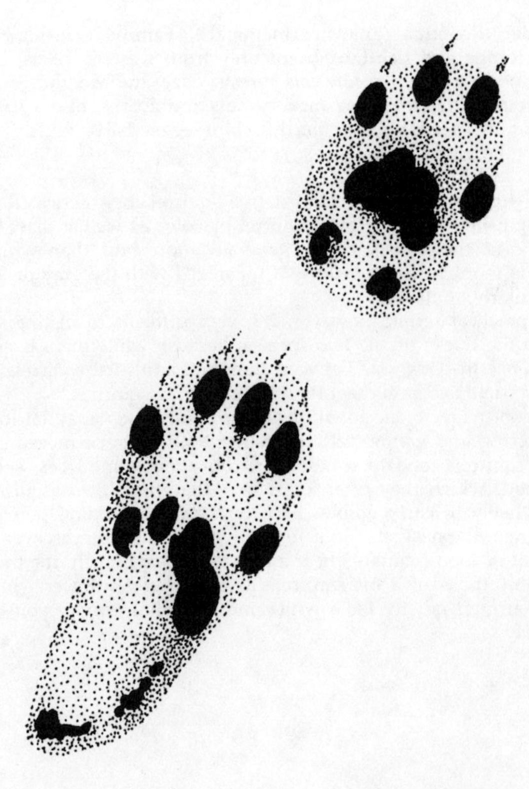

Gray Squirrel
life size in sand

FOX SQUIRREL *Sciurus niger*

Order: Rodentia (gnawing mammals). **Family:** Sciuridae (squirrels). **Range and habitat:** absent only from western Texas; in open hardwood or mixed coniferous forests. **Size and weight:** 25 inches; 2–3 pounds. **Diet:** prefers hickory nuts and acorns, also a variety of other nuts, seeds, fungi, berries, bird eggs, bark, buds. **Sounds:** ratchetlike, raspy barks.

The fox squirrel is the largest tree squirrel in America. It can be distinguished from the gray squirrel by color as well as size: the underside of the fox squirrel is generally more buff than white, and there is more rust and yellowish fur mixed with the gray of its coat, particularly on its tail.

In practical terms, however, it's very difficult to distinguish between the tracks of the two species, because while there is clearly a difference in body size between them, all adult fox squirrels do not necessarily have larger feet than all adult gray squirrels.

A good clue to the identity of the track maker may be found in wandering and eating patterns rather than in track measurements. Gray squirrels tend to wander far from their home trees, searching for food, which they often eat where they find it; fox squirrels are more likely to carry edibles back to a favorite feeding perch on or quite near the nest site. So if in following tracks, you discover a large deposit of food remnants near a log, stump, or branch, the tracks are probably those of a fox squirrel. To be certain, however, you could hang around quietly for a while and see what activity you can observe.

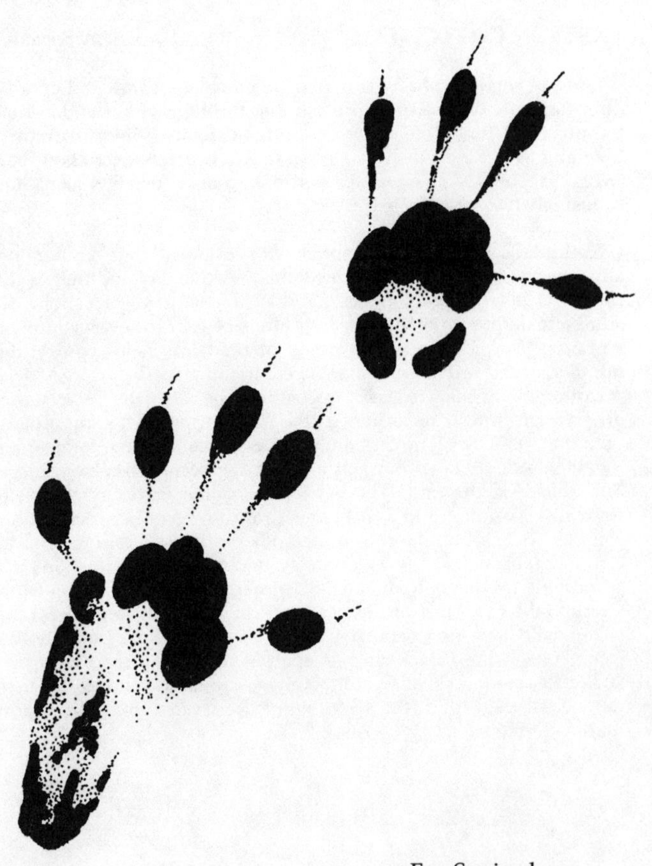

Fox Squirrel
life size in mud

EASTERN COTTONTAIL *Sylvilagus floridanus*

Order: Lagomorpha (rabbitlike mammals). **Family:** Leporidae (hares and rabbits). **Range and habitat:** throughout Texas; in brushy terrain, thickets, old fields, woods, often extending into urban areas. **Size and weight:** 13 inches; 3 pounds. **Diet:** green vegetation, bark, twigs, sagebrush, and juniper berries. **Sounds:** usually silent; loud squeal when extremely distressed.

Cottontails are the pudgy rabbits with cottonball tails, known to us all from childhood tales of Peter Rabbit. Active day and night, year-round, they're generally plentiful due in part to the fact that each adult female produces three or four litters of four to seven young rabbits each year. Of course, a variety of predators helps control their number, and few live more than a year in the wild.

Cottontail tracks are easily recognized because the basic pattern doesn't vary much, regardless of the rabbit's speed. It's important to note that, as with all rabbit family tracks, sometimes the forefeet land together, side by side, but just as often the second forefoot lands in line ahead of the first. The eastern cottontail leaves track clusters normally spanning 6 to 9 inches, with up to 3 feet between running clusters. Tracks like these in grasslands or creosote deserts of western and central Texas were probably made by a desert cottontail (*S. audubonii*); in the bottomlands of eastern Texas, the "cane-cutter" swamp rabbit (*S. aquaticus*) leaves identical or slightly larger tracks.

You will have no problem telling cottontail tracks from those of jackrabbits, whose track clusters span as much as 2 feet, with up to 20 feet separating clusters; and because jackrabbits tend to run up on the toes of their hind feet, they often leave *smaller* hind-foot imprints than cottontails do.

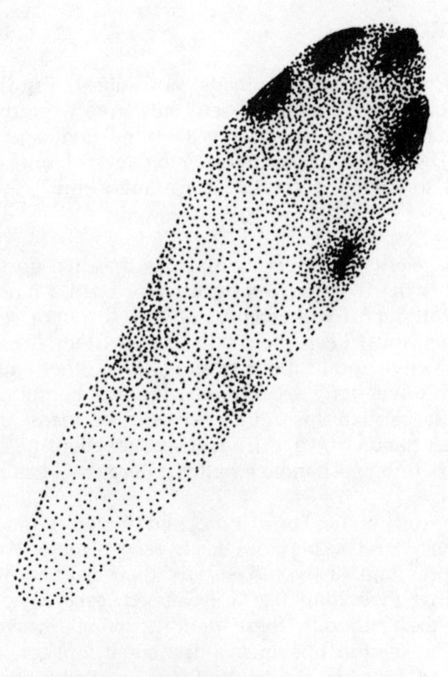

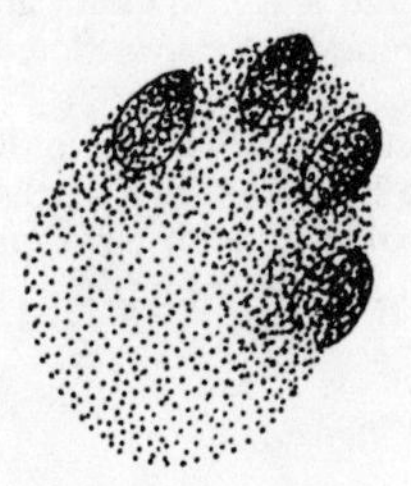

Eastern Cottontail
life size in sand

BULLFROG *Rana catesbiana*

Order: Salientia (frogs, toads, and allies). **Family:** Ranidae (true frogs). **Range and habitat:** absent only from western Texas; in ponds, lakes, marshes and swamps, year-round bodies of water. **Size and weight:** body 5–8 inches, with long legs; 4 ounces. **Diet:** insects. **Sounds:** low-pitched croaks, deep "jug-o-rum," especially at dawn and dusk.

Of the eighty-one species of frogs that live north of Mexico, the bullfrog is the largest. Varying in color from a mottled dark gray to green, with an off-white belly, bullfrogs live in or very near water because they must keep their skin wet and they breed in water. They use distinctive vocalizations to signal each other and to attract mates. The 4- to 6-inch-long vegetarian tadpoles take up to two years to develop into carnivorous adults. As with all toads and frogs, be sure that your hands are free from insect repellent or other caustic substances before you handle a bullfrog because it has particularly sensitive skin.

Frogs walk or hop in a more plantigrade manner than toads, so their tracks tend to be more easily recognizable. You will commonly find impressions of the full soles of their feet and might even be able to see that their hind feet are webbed, except the last joint of the longest toe, although these delicate membranes don't always imprint. The toed-in imprints of the small forefeet, combined with a straddle of 5 or 6 inches, should leave no doubt about the identity of bullfrog tracks, even if the hind-foot impressions are less than distinct.

Nine other frog species also live in various parts of Texas; all are significantly smaller than the bullfrog. Their tracks are similar to but smaller than those of the bullfrog, with correspondingly narrower straddle, shorter leaps, and shallower prints on a given surface.

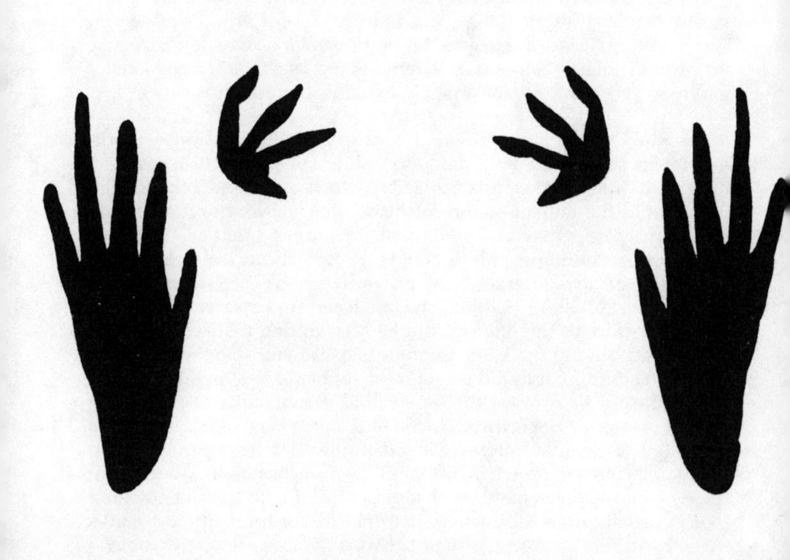

Bullfrog
life size in mud

MUSKRAT *Ondatra zibethicus*

Order: Rodentia (gnawing mammals). **Family:** Cricetidae (New World rats and mice). **Range and habitat:** absent only from central and southern Texas; in streams, lakes, ponds, and marshes. **Size and weight:** 24 inches; 4 pounds. **Diet:** aquatic vegetation; occasionally mollusks and small aquatic animals. **Sounds:** high-pitched squeaks.

The muskrat is a large brown rat with a volelike appearance, modified for its aquatic life by a rudderlike scaly tail and partially webbed hind feet. Muskrats associate readily with beavers and occasionally nest within the superstructure of beaver lodges. More often, muskrats burrow into riverbanks or construct lodges similar to those of beavers but extending only a couple of feet above water level and composed of aquatic vegetation, primarily grasses and reeds, rather than trees. Muskrat lodges always have underwater entrances. Mainly nocturnal, the muskrat can be seen during the late afternoon or at dusk, pulling the V of its ripple across a still water surface, tail sculling behind, mouth full of grass for supper or nest building.

Muskrat tracks are nearly always found in mud close to water. The muskrat is one of the few rodents with five toes on its forefeet, but its truncated inner toes often leave no imprint. It leaves tracks from about 2 inches apart (when walking) to 12 inches apart (when running), with the tail sometimes dragging as well. The track of the hind foot is usually more distinctive than that of the forefoot, and marks made by the stiff webbing of hair between the toes are often visible.

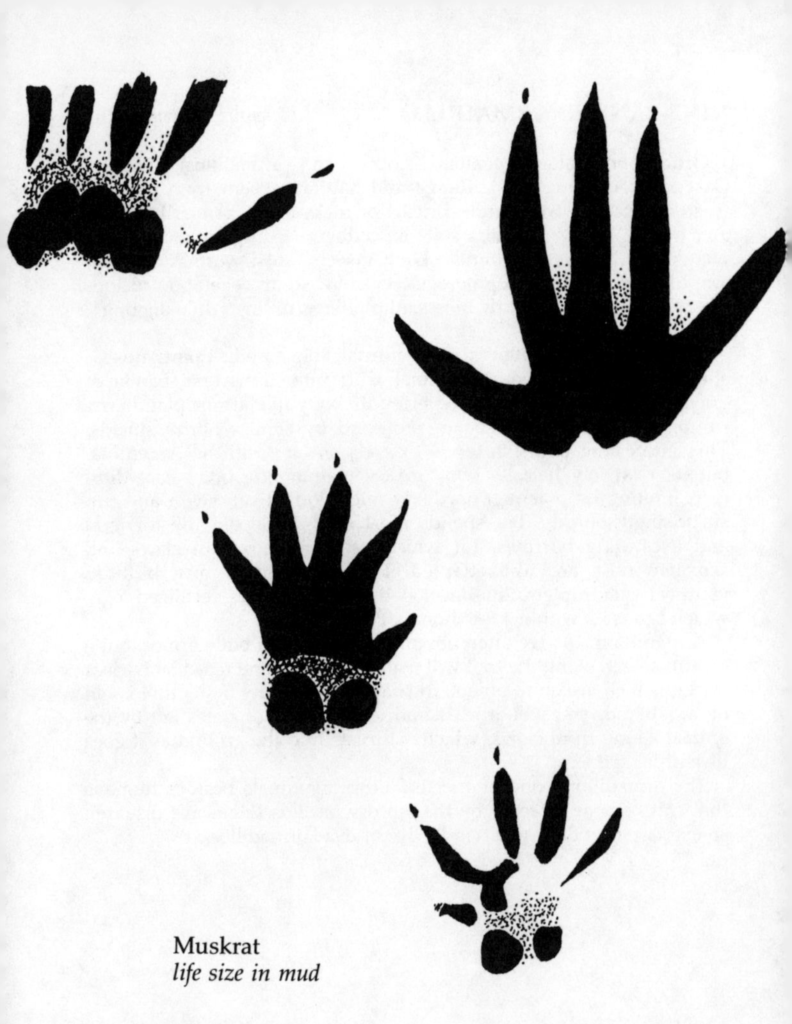

Muskrat
life size in mud

NINE-BANDED ARMADILLO *Dasypus novemcinctus*

Order: Edentata (anteaters, sloths, and armadillos). **Family:** Dasypodidae (armadillos). **Range and habitat:** absent from parts of western Texas; in woodlands, brushy or rocky areas, primarily in any area where sandy or moist soils encourage easy digging. **Size and weight:** 30 inches; 12 pounds. **Diet:** insects, ants, worms, crayfish, amphibians, bird and reptile eggs, carrion, some vegetation including berries. **Sounds:** nearly incessant piglike grunting while digging.

Our native armadillo is a nocturnal, big-eared, pointy-nosed, short-legged, heavy-bodied animal with nine transverse bands of gray, bony, segmented hide covering its body like armor plating. Its rump, shoulder, and head are protected by similar horny shields. This interesting animal's range extends as far south as Argentina, but its relatively hairless body makes maintaining body heat difficult, limiting its potential northerly migration. It can swim and run surprisingly quickly, but spends most of its time digging for food and excavating burrows, for which its large, powerful claws and sensitive nose are well adapted. Females annually give birth to identical quadruplets, all developed from a single fertilized egg, which can walk within a few hours of birth.

Armadillo tracks are often obscured by dragging body armor, but a careful search along the trail will usually reveal some reasonably clear tracks, which are quite unique in shape. Particularly if the trail is old or weathered, you will at least find telltale pairs of holes left by the animal's long front claws, which it thrusts into the ground as it goes its stiff-legged way.

The armadillo is one of the few other mammals besides humans that can become infected by the leprosy bacillus (Hansen's disease), so exercise caution if you handle live or dead armadillos.

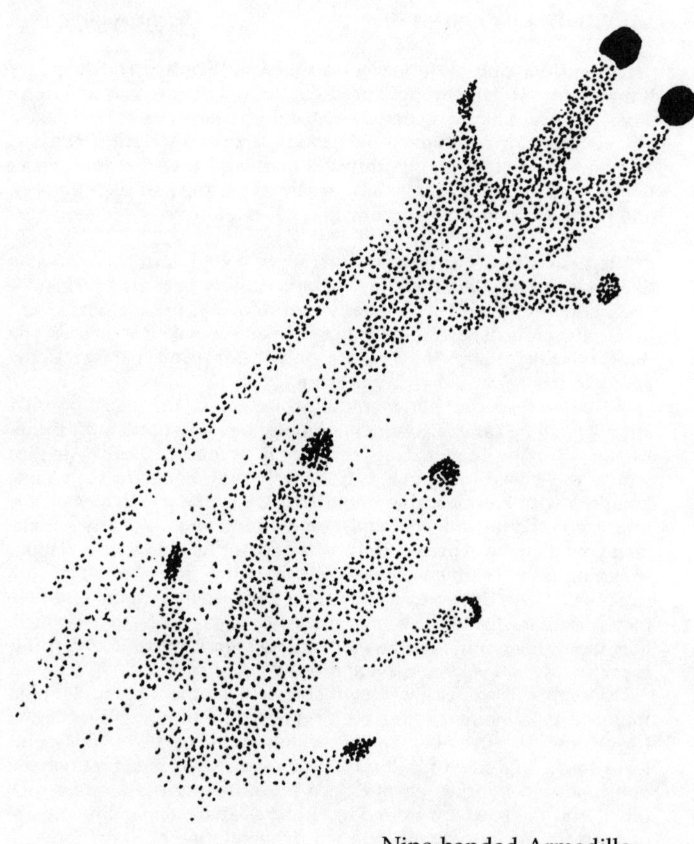

Nine-banded Armadillo
life size in sandy dirt

VIRGINIA OPOSSUM *Didelphis virginiana*

Order: Marsupialia (pouched mammals). **Family:** Didelphiidae. **Range and habitat:** throughout Texas; in woodlands and adjoining areas, and farmlands, generally remaining near streams and lakes; also common around human habitations. **Size and weight:** 25 inches; 12 pounds. **Diet:** this opportunistic omnivore prefers fruits, vegetables, insects, small mammals, birds, eggs, carrion; also garbage and pet food. **Sounds:** a gurgling hiss when annoyed.

The generally nocturnal opossum appears fairly ordinary: it looks like a large, long-haired rat, with pointed nose, pale-gray fur, and a long, scaly, reptilian tail. Primarily terrestrial, the opossum may nest in an abandoned burrow or a fallen tree, but will climb to escape danger. Other than climbing, its only defense mechanism is the ability to feign death, or "play possum."

In many respects, however, the opossum is the most peculiar animal residing on this continent. Among the oldest and most primitive of all living mammals, it is the only animal in North America with a prehensile (grasping) tail, the only nonprimate in the animal kingdom with an opposable (thumblike) digit (the inside toes on the hind feet), and the only marsupial on the continent. As many as fourteen young are born prematurely after only thirteen days of gestation, weighing only 1/15 of an ounce each (the whole litter would fit in a teaspoon!). The tiny babies crawl into their mother's pouch, where they remain for the next two months. After emerging from the pouch, they often ride around on the mother's back for some time. All of this is pretty unusual behavior, even in Texas.

Opossums leave easily identifiable tracks: the opposable hind thumb usually points 90 degrees or more away from the direction of travel, and the five toes spread widely. Like raccoons, opossums leave tracks in a row of pairs. Each pair consists of one forefoot and one hind-foot imprint, always close to or slightly overlapping each other, and the pairs are from 5 to 11 inches apart, depending on size and speed. The opossum's long tail frequently leaves drag marks on soft surfaces.

Virginia Opossum
life size in mud

GRAY FOX *Urocyon cinereoargenteus*
RED FOX *Vulpes vulpes*

Order: Carnivora (flesh-eating mammals). **Family:** Canidae (dogs). **Range and habitat:** gray fox, throughout Texas; red fox, in central and eastern Texas only; both in open forests and brushy, sparsely wooded fields and meadows nearby. **Size and weight:** 40–42 inches; 12–15 pounds. **Diet:** omnivorous, including small mammals, birds, insects, eggs, fruit, nuts, grains, and other forage. **Sounds:** gray fox, normally silent, occasionally short barking yips; red fox, a variety of doglike noises.

The gray fox is generally more nocturnal and secretive than the red fox. The only canine in America with the ability to climb, it frequently seeks refuge and food in trees, but cottontails are the mainstay of its diet. It typically dens among boulders on the slopes of rocky ridges or in rock piles, hollow logs, or the like; unlike the red fox, it uses these dens in winter as well as summer.

The sleek little red fox usually leaves a distinctive, nearly straight line of tracks, the forefoot track slightly wider than that of the hind foot. The claws always leave marks, although in deep snow the tail may brush over and obscure some of the finer points of the tracks. Red fox tracks could be mistaken for those of a small domestic dog, except that the fox's heel pad has a unique bar, and the heel pads of domestic dogs tend to be longer, extending forward between the outer toes. Also, a walking red fox leaves tracks from 12 to 18 inches apart, evidence of a somewhat longer stride than that of a similar-sized domestic dog.

Gray fox tracks are very similar to those of the red fox, except that the prints are usually more distinct due to the relative lack of fur on the animal's feet. The tracks always show the imprints of claws and may be the same size or slightly smaller and narrower than the red fox's, with 7 to 12 inches between walking prints. In open, sandy places in northern or western Texas, smaller tracks of similar shape may have been left by a kit or swift fox. These similarities necessitate field observations for positive identification.

Fox
life size in mud

BOBCAT
Wildcat

Felis rufus

Order: Carnivora (flesh-eating mammals). **Family:** Felidae (cats). **Range and habitat:** widespread throughout Texas; primarily in rimrock and chaparral or open woodlands, but adapts readily to swamp fringes, farmlands, and arid rocky or brushy areas. **Size and weight:** 30 inches; 35 pounds. **Diet:** small mammals and birds; rarely carrion. **Sounds:** capable of generic cat family range of noises.

The bobcat is a very adaptable feline, afield both day and night and wandering as far as 50 miles in a day of hunting, occasionally into suburban areas. It is primarily a ground hunter, but will climb trees and drop onto unsuspecting prey if the opportunity presents itself. You could mistake it for a large tabby cat with a bobbed tail, but the similarity ends there, for the bobcat has quite a wild disposition combined with greater size, strength, and razor-sharp claws and teeth.

You can expect to encounter bobcat tracks almost anywhere. You'll know the roundish tracks belong to a cat because the retractile claws never leave imprints and the toes usually spread a bit more than a dog's. Bobcat tracks are too large to be mistaken for those of a domestic cat, however. The animal's weight will have set the tracks deeper in a soft surface than you would expect from a house cat, and domestic cats have pads that are single-lobed at the front end. Bobcat tracks are clearly smaller than those of a mountain lion and are therefore easily identifiable by process of elimination in most areas.

Two similar-sized cat species, both endangered and relatively rare, overlap the bobcat's range in south and central through eastern Texas, leaving tracks quite similar to the bobcat's. See Ocelot and Jaguarundi for specific differences.

Bobcat
life size in mud

OCELOT *Felis pardalis*
JAGUARUNDI *Felis yagouaroundi*

Order: Carnivora (flesh-eating mammals). **Family:** Felidae (cats). **Range and habitat:** ocelot absent only from northern and western Texas, jaguarundi inhabits extreme southern Texas only; the ocelot prefers forested and brushy areas, jaguarundi prefers thickets, especially with spiny plants. **Size and weight:** ocelot, 30 inches, 30 pounds; jaguarundi, 25 inches, 16 pounds. **Diet:** both eat a variety of small mammals, birds, and reptiles, possibly also fruit, fish, carrion when available. **Sounds:** generally quiet but capable of generic cat family range of noises.

The ocelot looks like a bobcat with the markings of a jaguar: black-bordered brown spots, some of which merge into elongated lines. The jaguarundi is shaped more like a large domestic cat, with shorter legs than bobcats and ocelots; it occurs in one of three colors: dark gray to black, paler gray, or rust to buff. Both of these cats, like other North American cats, are generally secretive and solitary, hunting on the ground (although they do climb and swim well), usually at night but also infrequently during daylight hours.

Both the jaguarundi and ocelot are more common from Mexico south throughout Central and South America; they occur only rarely north of the Rio Grande, the ocelot ranging farther north than the jaguarundi. Both are considered endangered species in the United States.

Both of these cats overlap the range of the bobcat, and their tracks all have the same cat family characteristics. The jaguarundi's tracks are usually elongated and narrow, less than 1½ inches wide, with strides of 7 to 10 inches. The bobcat and ocelot, both longer legged, leave walking prints 10 to 16 inches apart, with leaps of 4 to 8 feet; and their feet are larger. Bobcat tracks are usually about 2 inches long and wide; ocelot tracks may be from 2 to 2½ inches wide, and they lack the scalloped heel pad shape of the bobcat. In most tracking conditions these differences may not be discernible, in which case the odds clearly favor the bobcat, by far the more common and numerous wild Texas cat.

Ocelot
life size in dirt

COYOTE *Canis latrans*
Brush wolf, prairie wolf

Order: Carnivora (flesh-eating mammals). **Family:** Canidae (dogs). **Range and habitat:** widespread throughout Texas; primarily in prairies, open woodlands, and brushy fringes, but very adaptable; can turn up anywhere. **Size and weight:** 48 inches; 45 pounds. **Diet:** omnivorous, including rodents and other small mammals, fish, carrion, insects, berries, grains, nuts, and vegetation. **Sounds:** wide range of canine sounds; most often heard yelping in group chorus late at night.

An important controller of small rodents, the smart, adaptable coyote is—unlike the gray wolf—steadily expanding its range. About the size of a collie, the coyote is a good runner and swimmer and has great stamina. Despite its wide range, it is shy, and you will be lucky to see one in the wild.

Typically canine, the coyote's forepaw is slightly larger than the hind paw, and the toes of the forepaw tend to spread wider, though not as wide as the bobcat's. The toenails nearly always leave imprints. The shape of coyote pads is unique, the pads of the forefeet differing markedly from those of the hind feet, as shown. Also, the outer toes are usually slightly larger than the inner toes on each foot. The coyote tends to walk in a straight line and keep its tail down, which often leaves an imprint in deep snow. These characteristics plus walking strides of 8 to 16 inches and leaps to 10 feet may help you distinguish coyote tracks from those of domestic dogs with feet of the same size.

Coyote
life size in mud

COATI *Nasua nasua*

Order: Carnivora (flesh-eating mammals). **Family:** Procyonidae (raccoons, ringtails, and coatis). **Range and habitat:** from Alpine southeast to the Gulf of Mexico, within 30 miles of the Rio Grande; in rocky, wooded canyons and uplands, open forests, often near water. **Size and weight:** 40 inches (half of which is tail); 25 pounds. **Diet:** omnivorous, prefers fruit, also eats nearly all small animals, birds, reptiles, or insects. **Sounds:** in groups, coatis often snort, grunt, chatter, whine, or scream.

The coati looks like a raccoon with a long, faintly banded furry tail, a dark face with white spots above and below its eyes, and an elongated nose. A tropical jungle animal, it has expanded its range northward from South and Central America. The coati is active during daylight hours, and is a gregarious, social animal, often traveling in noisy, boisterous groups, which may consist of a dozen or more members. Alone or in groups, the coati is a fine climber and a good swimmer. Not a particularly timid animal in general, it becomes downright friendly with people who give it the slightest encouragement.

Coati tracks are rare, but fairly easy to identify. The coati digs more than the raccoon, so its claws tend to be longer and its toes more muscular and blunt, while the raccoon's toes are longer and thinner, with shorter claws. It is also less flatfooted than its cousin the raccoon, and you will find its tracks in mud less often than the raccoon's because the coati does not wash its food. The coati's tracks are also very similar to, although smaller than, most badger tracks, but the badger leaves many more signs of extensive earthmoving in the vicinity. Finally, the coati is the only small carnivore in North America that habitually travels in large groups.

Coati
life size in mud

RACCOON

Procyon lotor

Order: Carnivora (flesh-eating mammals). **Family:** Procyonidae (raccoons, ringtails, and coatis). **Range and habitat:** throughout Texas; varied, mostly in forest fringes and rocky areas near streams, ponds, and lakes. **Size and weight:** 36 inches; 25 pounds. **Diet:** omnivorous, including fish, amphibians, mollusks, insects, birds, eggs, mice, carrion, berries, nuts, and vegetation. **Sounds:** a variety of shrill cries, whistles, churrs, growls, and screeches.

From childhood most of us know the raccoon by its mask of black fur and its black tail stripes on an otherwise grayish-brown body. It's familiar as a character in kids' books and frontier lore, frequently seen as a road kill, and is both curious and bold enough to be a fairly common visitor to campgrounds and even residential homes nearly everywhere within its range. Chiefly nocturnal, raccoons are commonly sighted in suburban neighborhoods, raiding garbage cans and terrorizing family hounds. Interesting and intelligent animals with manual dexterity of great renown, raccoons are also reputed to make lively and intriguing pets, provided they are closely supervised.

Raccoons like to wash or tear food items apart in water, which apparently improves their manual sensitivity. Much of their food comes from aquatic prospecting, so you will often find their tracks near water. When a raccoon walks, its left hind foot is placed next to the right forefoot, and so forth, forming paired track clusters. Running-track clusters tend to be bunched irregularly. The walking stride of a raccoon is about 7 inches; leaps average 20 inches.

The ringtail, which lives in rocky habitats around most of the region, could be mistaken for a raccoon at a quick glimpse. The ringtail's body is more foxlike, its face has white eye rings rather than a black mask, and its tail has black *and white* stripes. Interestingly, its tracks resemble weasel family tracks without claw prints, rather than those of its raccoon relatives.

Raccoon
life size in mud

BADGER *Taxidea taxus*

Order: Carnivora (flesh-eating mammals). **Family:** Mustelidae (the weasel family). **Range and habitat:** widespread throughout Texas; in treeless meadows, semiopen prairies, grasslands, and deserts at all altitudes, wherever ground-dwelling rodents are abundant. **Size and weight:** 28 inches; 20 pounds. **Diet:** carnivorous, including all small rodents, snakes, birds, eggs, insects, and carrion. **Sounds:** may snarl or hiss when alarmed or annoyed.

The badger is a solitary creature that digs up most of its food and tunnels into the earth to escape danger. Its powerful, short legs and long, strong claws are well suited to its earthmoving ways, and badgers can reputedly dig faster than a man with a shovel. Above ground the badger is a fierce fighter threatened only by much larger carnivores. The animal is active in daylight and not too shy, even entering campgrounds in its search for food. Its facial markings are quite distinctive and easily recognized: the face is black with white ears and cheeks and a white stripe running from its nose over the top of its head. The rest of the body is light brown or gray, and the feet are black.

Badger tracks show five long, clear toe prints of each foot and obvious marks left by the long front claws. The animal walks on its soles, which may or may not leave complete prints. Its pigeon-toed trail may be confused with the porcupine's in deep snow, but a porcupine trail will invariably lead to a tree or into a natural den, a badger's to a burrow of its own excavation. Another clue: the badger's short, soft tail rarely leaves a mark.

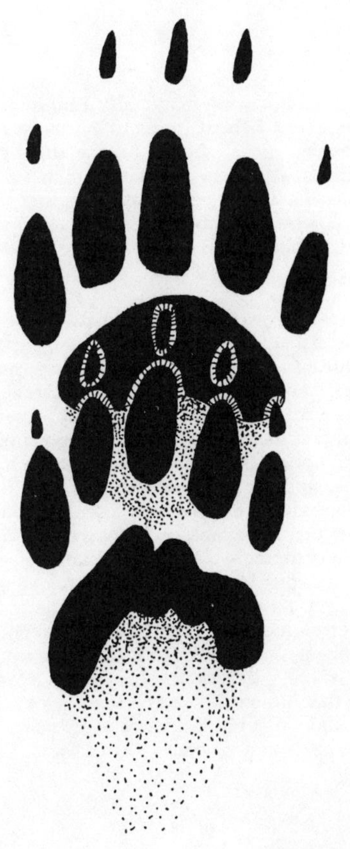

Badger
life size in mud

PORCUPINE
Porky, quill pig

Erethizon dorsatum

Order: Rodentia (gnawing mammals). **Family:** Erethizontidae (porcupines). **Range and habitat:** parts of northern and western Texas; usually in forested areas, also in brushy fringes, fields, meadows, and semidesert areas; a very adaptable animal. **Size and weight:** 30 inches; 25 pounds. **Diet:** vegetarian, including bark, leaves, fruits, berries, nuts, flowers. **Sounds:** normally quiet; capable of a great variety of grunts, whines, and harmonicalike noises and rapid teeth clicking.

The porcupine is one of the few animals whose tracks you can follow with reasonable expectation of catching up with their maker. Often out during daylight hours, it moves quite slowly if not alarmed, stops frequently to nibble at vegetation, and does not see well; so if you're quiet, you can usually observe this peaceable animal at your and its leisure. An alarmed porcupine climbs a tree to escape danger, only using its quills as a last-ditch defense against an outright attack; and the porcupine cannot fling its quills, so there's no danger to any creature with enough sense to stay out of direct contact, a requisite that regrettably excludes many domestic dogs.

Often the porcupine's distinctive shuffling gait and dragging whisk-broom tail may be the only clear track signs it leaves behind, especially in deep snow. In winter, porcupine trails often lead to or away from a coniferous tree, where the animal both sleeps and dines on bark and needles; alternatively, it may hole up in a den beneath a stump or in another ground-level shelter. Occasionally a piece of snow or mud that has stuck to a porcupine's foot will dislodge intact, revealing the unique pebbled texture of its soles. Imprints from the long claws are also often visible.

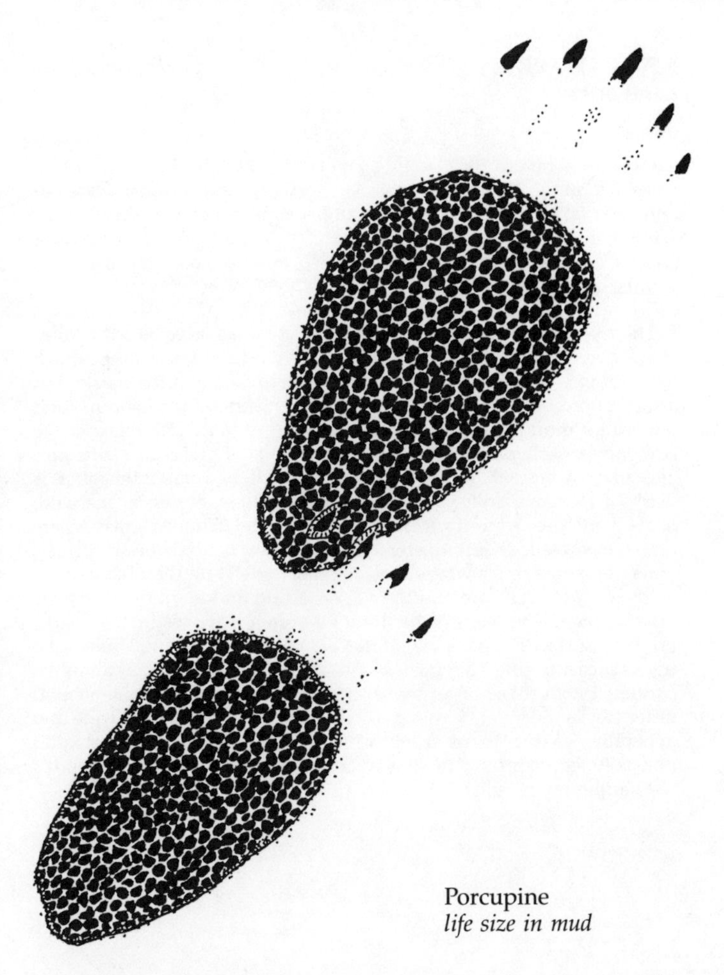

Porcupine
life size in mud

RIVER OTTER *Lutra canadensis*
Land otter

Order: Carnivora (flesh-eating mammals). **Family:** Mustelidae (the weasel family). **Range and habitat:** northern and eastern Texas; in and near lakes and streams. **Size and weight:** 48 inches; 25 pounds. **Diet:** fish, amphibians, mollusks and other aquatic invertebrates, snakes, turtles, birds, eggs. **Sounds:** chirps, chatters, chuckles, grunts, and growls.

The river otter is a dark brown weasel about as large as a medium-sized dog, with a thick, hairless tail adapted for swimming, much like that of a muskrat; in fact, the river otter resembles the muskrat in appearance and habitat, but is much larger, strictly carnivorous, and quite a bit more animated. Both in and out of water, alone or in the company of others, the river otter seems to be a graceful, exuberant, and playful animal. Active during the daylight hours, the otter is wary of humans. Still, you might occasionally sight one in the wild; more commonly you may find, in summer, the flattened grass where otters have rolled, leaving their musky odor behind; or in winter, marks on snow or ice where they've playfully slid on their bellies.

River otter tracks are relatively easy to find and identify within the otter's range. The webs of the hind feet often leave distinctive marks on soft surfaces, and claw marks usually are present. Individual tracks measure up to 3½ inches across and due to their size cannot be confused with those of any other animal with similar aquatic habitat. Otter tracks meander, forming a trail roughly 8 to 10 inches wide and generally leading to or from water systems. Also, the river otter normally leaves groups of four tracks 13 to 30 inches apart, when it's not sliding on its belly.

River Otter
life size in mud

COLLARED PECCARY *Dicotyles tajacu*
Javelina

Order: Artiodactyla (even-toed hoofed animals). **Family:** Tayassuidae (New World swine). **Range and habitat:** absent from northern and eastern Texas; in brushy, semiarid desert, chaparral, mesquite, rocky canyons, along cliffs, and around waterholes. **Size and weight:** 36 inches in length, 24 inches to shoulder; 45 pounds. **Diet:** primarily herbivorous, including cacti, mesquite, fruit, sotol, lechuguilla; also nuts, berries, grubs, bird eggs where available. **Sounds:** grunts when alarmed.

This small, dark gray pig with its thin collar of white hair is descended from larger wild pigs that lived in North America about 25 million years ago. The collared peccary is primarily a tropical animal, its range extending south from the American Southwest to South America. Peccaries are most active in morning and late afternoon, when you might see herds of up to thirty animals wandering around, eating prickly pear or scrub oak acorns and grunting softly together. They are good swimmers and can run at speeds of up to 25 miles per hour when necessary.

Diminutive collared peccary tracks are very difficult to confuse with any others. A very young deer leaves similar tracks, but infant deer don't travel in herds and you will always find the larger tracks of adult deer near young deer tracks. Wild boar tracks are similar also, but almost twice as large, with dewclaw imprints nearly always present and a walking stride of about 18 inches, compared with 10 inches or less for an adult peccary.

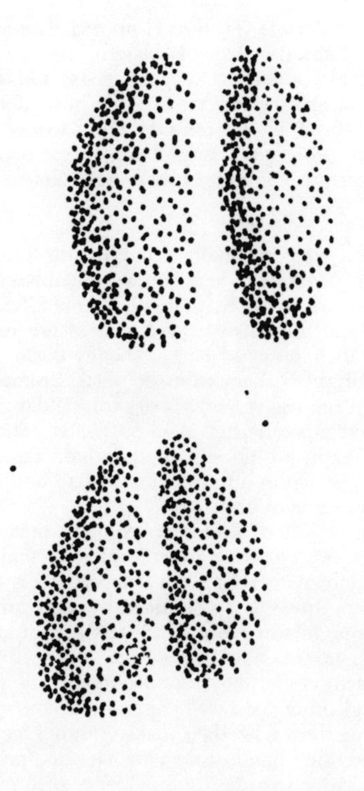

Collared Peccary
life size in sand

WILD BOAR *Sus scrofa*

Order: Artiodactyla (even-toed hoofed mammals). **Family:** Suidae. **Range and habitat:** parts of eastern, central, and western Texas; habitat variable, including dense forests, adjoining brushlands, dry hills, and swampy fringes, particularly in winter. **Height and weight:** 36 inches at shoulder; 300 pounds. **Diet:** acorns and other nuts, roots, grasses, fruit, also small amphibians, eggs, occasionally small mammals and carrion. **Sounds:** grunts and squeals typical of domestic pigs.

Wild boars were originally brought from Europe to stock hunting preserves in America, where some inevitably escaped and bred with feral domestic pigs; today, the wild boars of Texas are almost entirely hybrids. These brown or gray creatures are recognizably piglike in shape, but their humped backs, shaggy coats, and formidable tusks readily distinguish them from domestic animals. Strong, agile, and occasionally aggressive, wild boars can be dangerous to encounter in the dense vegetation they tend to prefer. They are most active at dawn and dusk, are good swimmers and fast runners, and usually move about in family groups of six or so animals, occasionally congregating in herds of up to fifty.

The cloven tracks of wild boars could be mistaken for those of mule deer, which overlap their range, but note that wild boar tracks are more rounded, with toes more splayed, heels together—often to the extent that no break is apparent in the imprint at all—and that dewclaw imprints, in soft ground, are farther out to the sides than you'd find in deer tracks. Wild boars have strides of about 18 inches, usually in a narrow line; they root around on the ground in their search for nuts and other food, leaving behind characteristic diggings, and rub or gouge trees with their tusks within 3 feet of the ground. Mule deer, on the other hand, leave narrower hoofprints with heels usually separated, 24-inch strides, and evidence of browsing in foliage 5 or 6 feet above the ground.

Wild Boar
life size in mud

PRONGHORN
Antelope

Antilocapra americana

Order: Artiodactyla (even-toed hoofed mammals). **Family:** Antilocapridae (pronghorns). **Range and habitat:** western half of Texas; on open prairies and sagebrush plains. **Height and weight:** 36 inches at shoulder; 120 pounds. **Diet:** vegetation, including weeds, shrubs, grasses, and herbs; fond of sagebrush. **Sounds:** usually silent, but capable of a loud whistling sound when startled.

The pronghorn is the most easily observed wild hoofed mammal in Texas. This sociable animal lives in flat, open country, and its tan hide marked with striking rump patches, white underside, and facial spots is easy to recognize. The fastest animals in North America, pronghorns race around the plains en masse at speeds of 45 miles per hour or so.

Because the pronghorn is easily sighted from a distance, you normally won't have to rely on tracks for identification, but the tracks should be easy to recognize because, unlike the mule deer, which shares much of its range, the pronghorn has no dewclaws. It is more gregarious than deer, and its running characteristics are unique: a large group of pronghorns may run for a mile or more in a straight line, whereas deer tend to run only when startled and then they don't run very far. The pronghorn's great speed produces an average separation of 14 feet between track clusters; running white-tailed deer average 6 feet, and mule deer average about 10 feet.

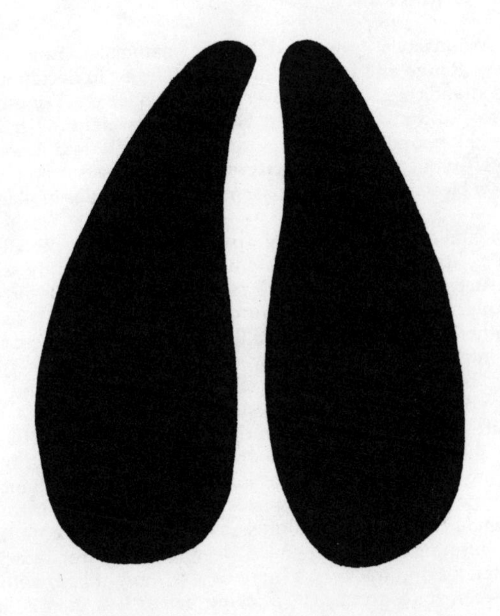

Pronghorn
life size in mud

WHITE-TAILED DEER *Odocoileus virginianus*
Whitetail, Virginia deer

Order: Artiodactyla (even-toed hoofed mammals). **Family:** Cervidae (deer). **Range and habitat:** throughout Texas; in deciduous and mixed woodlands and nearby river bottomlands, creeksides, open brushy areas, and swamp fringes. **Height and weight:** 42 inches at shoulder; 250 pounds. **Diet:** browse from shrubs and lower tree limbs; less frequently fungi, nuts, grains, grasses, and herbs. **Sounds:** low bleats, guttural grunts, snorts, and whistles of alarm.

The white-tailed deer is recognizable in the wild by the all-white underside of its tail, which the animal raises prominently when it runs. It usually has a home range of only a square mile or so, although some migrate to swamps in cold weather. White-tailed deer usually gather in groups of no more than three animals, except in the dead of winter when the group may swell to twenty-five. They spend their days browsing and quietly chewing their cuds and when startled run only short distances to the nearest cover.

The white-tailed deer's range overlaps those of several other deer species in Texas, but its tracks are easy enough to identify. Individual tracks are relatively long and slender, averaging about 3 inches in length; the dewclaws will leave prints in snow or soft earth. Taken alone, a single track could be confused with one left by a mule deer, but when running, the white-tailed deer tends to trot rather than bound, often leaving tracks in a more or less straight line, with up to 6 feet between track groups. Its walking gait is less than 20 inches, with tracks frequently doubled up, as the hind feet cover prints left by the forefeet. Also, the white-tailed deer generally lives in open forests and upland meadows, whereas the mule deer prefers either more mountainous terrain or open prairies and desert plains.

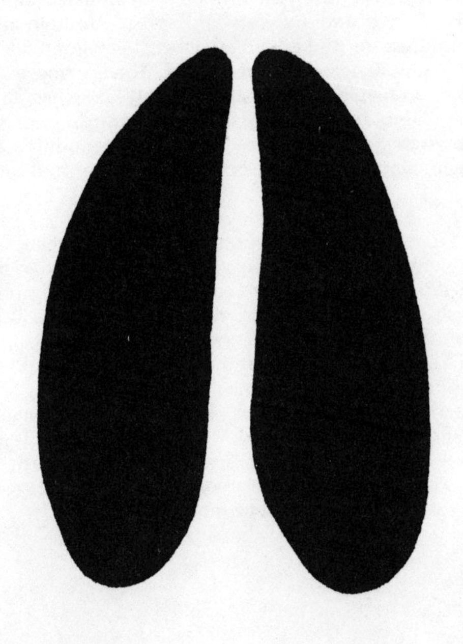

White-tailed Deer
life size in mud

MULE DEER *Odocoileus hemionus*

Order: Artiodactyla (even-toed hoofed mammals). **Family:** Cervidae (deer). **Range and habitat:** widespread throughout western Texas; in mixed forests and adjoining areas at all elevations, regularly moving into grasslands, chaparral, and desert fringes that have browse plants available. **Height and weight:** 42 inches at shoulder; 350 pounds. **Diet:** herbivorous, including leaves, grasses, grains, nuts, and berries. **Sounds:** generally silent, but produces a snorting whistle when alarmed; makes occasional grunts and other vocal noises.

The mule deer is easily recognized by its large ears and black-tipped tail, which it holds down when running. Active during the day and at dusk, the mule deer can be observed fairly often in the wild. Skittish, at the first sign of alarm it flees with a unique feet-together bounding gait, all four hooves landing and taking off at the same time. This gait is an adaptation to life in rugged, often densely brushy terrain. The mule deer is also a strong swimmer.

Mule deer tracks show small, slender hooves usually spread slightly at the heels, often with dewclaw impressions. The doubled-over walking tracks are usually less than 2 feet apart; at speeds beyond a walk, the mule deer bounds, leaving very distinctive clusters of parallel tracks 10 feet or more apart.

Mule Deer
life size in mud

BIGHORN SHEEP *Ovis canadensis*
Desert bighorn, mountain sheep

Order: Artiodactyla (even-toed hoofed mammals). **Family:** Bovidae (cattle, sheep, and goats). **Range and habitat:** western Texas; in mountainous, sparsely populated terrain and high, hilly desert, avoiding forested areas. **Height and weight:** 45 inches at shoulder; 275 pounds. **Diet:** variety of high-altitude browse, grasses, herbs, lichens. **Sounds:** loudest and most recognizable are the sounds of head and horn butting made by competing rams during late fall; coughs, grunts, bleating.

This big sheep, with its distinctive white rump and heavy coiled horns, lives in remote areas of western Texas, where its hooves with hard outer edges and spongy centers give it excellent agility on rocky surfaces. It tends, however, to prefer high mountain meadows and scree slopes. In the summer, you commonly see groups of about ten ewes and lambs grazing or lying around chewing their cuds. In winter, rams join the herd, which may grow in size to 100 animals or more. The most distinctive behavior of the bighorn sheep is the frenzied, high-speed head butting engaged in by competing rams before the autumn mating season; noise from the impact carries a great distance in the open mountain country. Thanks to nature documentaries shown on television, people have had the opportunity to witness this head butting ritual without having to venture into remote mountainous wildlands, thus saving wear and tear on fragile ecosystems and human knee joints.

Because of the habitat where they are commonly found, bighorn sheep tracks are rarely confused with those of other hoofed animals. The hooves average 3½ inches in length; rather blunt and square, they may show signs of wear and tear from the scree slopes; and dewclaw prints are never left.

Bighorn Sheep
life size in mud

BLACK-TAILED JACKRABBIT *Lepus californicus*
Jackass rabbit

Order: Lagomorpha (rabbitlike mammals). **Family:** Leporidae (hares and rabbits). **Range and habitat:** throughout Texas; in open prairies and sparsely vegetated sage and cactus country. **Size and weight:** 20 inches; 6 pounds. **Diet:** mostly grasses and other green vegetation, often along highway edges; also shrubs, buds, bark, twigs, and cultivated crops. **Sounds:** normally silent.

The black-tailed jackrabbit is easily recognized by its year-round light gray fur, white underside, and distinctive black fur patch on top of its tail. It is most active from dusk to dawn and spends most of its days lying in depressions it scoops out at the base of a bush, by a rock, or at any other spot that gives it a bit of protection. These jackrabbits are sociable and are often seen feeding in small groups.

Jackrabbit tracks are usually easy to identify. Typical of the rabbit family, the hind feet land ahead of the forefeet, leaving the familiar four-print clusters. The two tracks on the left side of the illustration show flatfooted walking tracks. As a jackrabbit's speed increases, its heels lift until, at top speed of 35 to 40 miles per hour, only the toes leave marks, which sometimes resemble coyote tracks, as illustrated on the right. The most distinctive track characteristic of jackrabbits, however, is that at running speed they leap repeatedly from 7 to 12 feet and more. Coyotes are capable of 10-foot bounds, but normally average about 6 feet. The cottontail rabbits of Texas bound no more than 3 feet per jump.

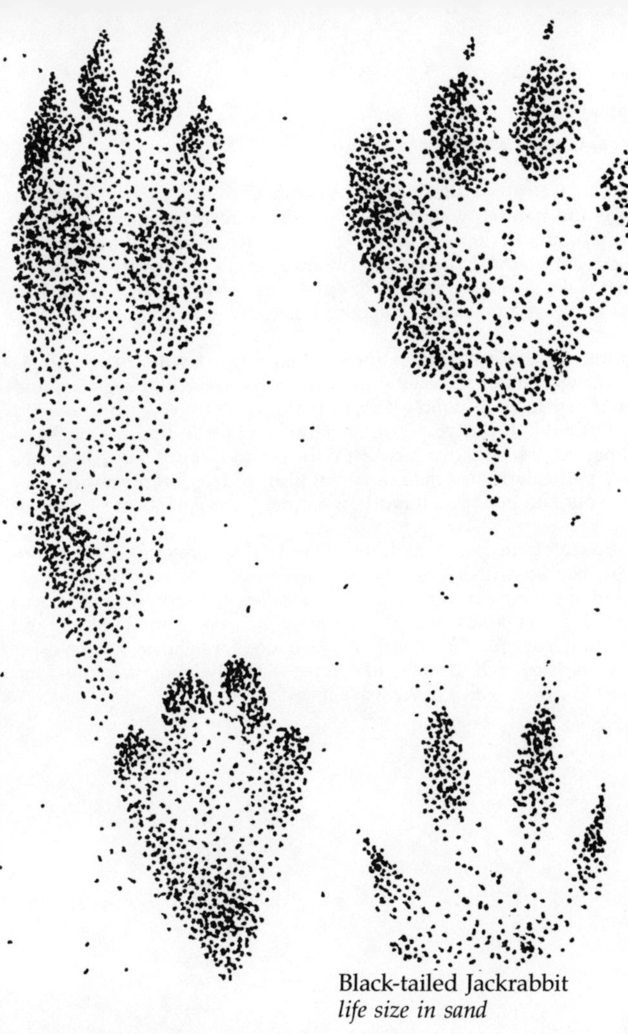

Black-tailed Jackrabbit
life size in sand

MOUNTAIN LION *Felis concolor*
Puma, cougar, panther, catamount

Order: Carnivora (flesh-eating mammals). **Family:** Felidae (cats). **Range and habitat:** widespread throughout Texas; in rugged wilderness mountains, forests, and swamp fringes. **Size and weight:** 84 inches; 200 pounds. **Diet:** primarily deer, small mammals, and birds; occasionally domestic animals. **Sounds:** generally quiet, but capable of a variety of loud feline screams, hisses, and growls.

Hunted to the verge of extinction, our large, tawny, native American cat with its long, waving tail is now so scarce and secretive and is confined to such remote terrain that you'd be very lucky to sight one in the wild. But you at least have a chance to find the tracks of this big cat, which hunts mostly on the ground. It occasionally climbs trees, particularly to evade dogs but also to drop onto unwary prey (the mountain lion is an important natural control of the deer population).

Because of their size, mountain lion tracks normally can't be confused with any others. Spacing of tracks will also be what you'd expect of the large cat: trail width of 12 inches or more, walking tracks over 20 inches apart, 3 feet separating pairs of loping tracks, and bounding leaps of 12 feet or more. Also look for tail drag marks, especially in snow. Of course, like most cats, the mountain lion has retractile claws, which never leave marks.

Mountain Lion
life size in mud

RED WOLF *Canis rufus*

Order: Carnivora (flesh-eating mammals). **Family:** Canidae (dogs). **Range and habitat:** southeastern Texas; in forested or brushy areas, coastal plains, swamps and bayous. **Size and weight:** 60 inches; 65 pounds. **Diet:** primarily rabbits, also small or young animals, birds, and crabs along the Gulf Coast. **Sounds:** variety of dog family noises and music, often a clear, high tremolo chorus on moonlit nights.

The red wolf usually looks and acts like a large coyote with a finely mottled grayish red coat, although black, brown, yellowish, and gray specimens have also been seen; a great variety of size and color can be expected, because the red wolf interbreeds readily with the coyote. In fact, since the red wolf's range is wholly overlapped by the coyote, this uncommon Texas wolf may be doomed to exterminate itself through hybridization. Like the coyote, the red wolf is largely nocturnal and, although little is actually known of its habits, it is said to run with its tail up, whereas the coyote keeps its tail down. With patience and the mouse-squeak made by kissing the back of your hand, you can sometimes lure canines in for precise identification.

Without an actual sighting, it will be difficult to positively identify red wolf tracks because, well, who can say if the tracks were left by a large coyote, a small red wolf, or any of numerous hybrid offspring? Generally speaking, the red wolf's tracks are larger than the coyote's and the outer toes are slightly smaller than the middle toes. If track length is longer than 3 inches, and there are no tail drag marks on soft surfaces, the trail is likely that of the elusive red wolf.

Red Wolf
½ life size in mud

BEAVER *Castor canadensis*

Order: Rodentia (gnawing mammals). **Family:** Castoridae (beavers). **Range and habitat:** throughout Texas; in streams and lakes with brush and trees or in open forest along riverbanks. **Size and weight:** 36 inches; 55 pounds. **Diet:** aquatic plants, bark, and the twigs and leaves of many shrubs and trees, preferably alder, cottonwood, and willow. **Sounds:** nonvocal, but smacks tail on water surface quite loudly to signal danger.

This industrious aquatic mammal is the largest North American rodent. Although it sometimes lives unobtrusively in a riverbank, usually it constructs the familiar beaver lodge: a roughly conical pile of brush, stones, and mud extending as much as 6 feet above the surface of a pond. It gnaws down dozens of small softwood trees, with which it constructs a conspicuous system of dams often several hundred yards long. A beaver can grasp objects with its forepaws and stand and walk upright on its hind feet. It uses its flat, scaly, strong tail for support out of water and as a rudder when swimming. Gregarious animals, beavers work well together on their collective projects. They are active day and night year-round, but may operate unobserved beneath the ice during much of the winter, using subsurface lodge entrances.

If you are lucky, the large, webbed hind-foot tracks left by a beaver will be clear, with 6 to 8 inches between pairs. Beavers frequently, however, obscure part or most of their tracks by dragging their tails and/or branches over them, leaving a trail much like that of a 6-inch-wide turtle, except that the beaver's tail drag trail zigzags slightly every 6 or 8 inches. The zigzag is the key to identification; a turtle moves in reasonably long, straight segments until it changes direction significantly.

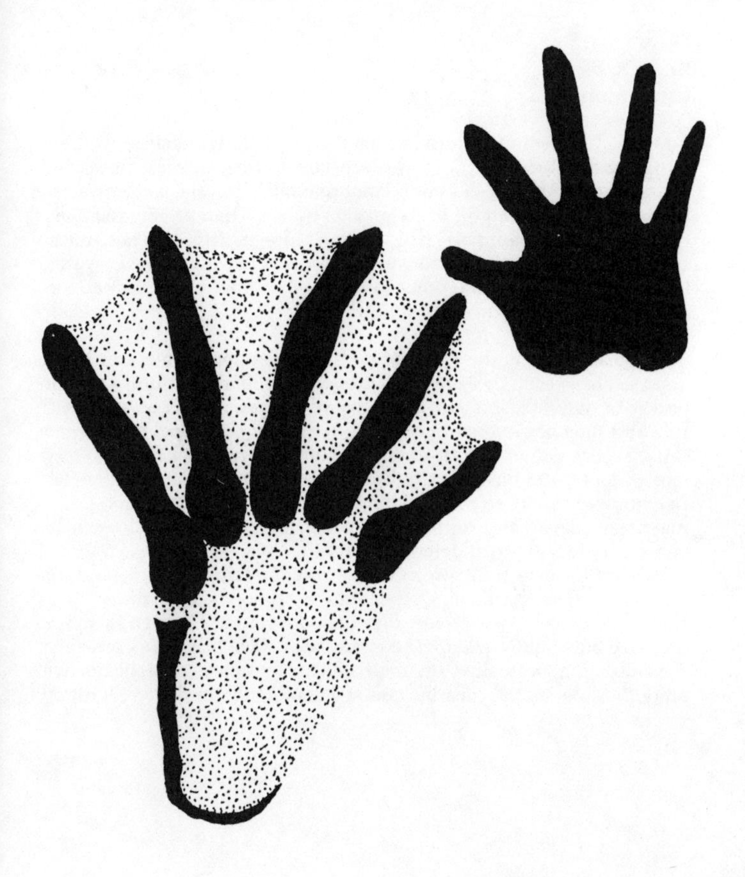

Beaver
½ life size in mud

BLACK BEAR *Ursus americanus*
Cinnamon bear

Order: Carnivora (flesh-eating mammals). **Family:** Ursidae (bears). **Range and habitat:** parts of western and northern Texas; primarily in medium-to-higher-elevation mountainous forests and swamp fringes. **Size and weight:** 6½ feet; 450 pounds. **Diet:** omnivorous, including smaller mammals, fish, carrion, insects, fruit, berries, nuts, and succulent plants. **Sounds:** usually silent, but may growl, grunt, woof, whimper, or make other immediately recognizable indications of annoyance or alarm.

The black bear is the smallest and most common American bear. You may have seen these animals around rural garbage dumps and in parks. In the wild, it is shy and wary of human contact as a general rule and thus not frequently sighted. If you do sight one, however, it can be very dangerous to underestimate it. The black bear is very strong, agile, and quick. It climbs trees, swims well, can run 25 miles per hour for short stretches, and above all else, is unpredictable; it may seem docile and harmless in parks, but it has been known to chase people with great determination.

Be alert for bear trails, worn deep by generations of bears, and for trees with claw marks and other indications of bear territory. Bear tracks are usually easy to identify; they are roughly human in shape and size but slightly wider. The large claws leave prints wherever the toes do. If a bear slips on mud or ice, its soles leave distinctive smooth slide marks; nearby you will no doubt find more orderly tracks.

Black Bear
½ life size in mud

Birds

KILLDEER *Charadrius vociferus*

Order: Charadriiformes (shorebirds, gulls, and terns). **Family:** Charadriidae (plovers). **Range and habitat:** year-round resident in most of Texas; common along shorelines, also found on inland fields and pastures. **Size and weight:** length 10 inches, wingspan 12 inches; 4 ounces. **Diet:** insects and larvae, earthworms, seeds. **Sounds:** repeats its name as its call.

The killdeer is one of the most common and recognizable shorebirds. Its two black breast bands are distinctive, as is its habit of feigning injury to lead intruders away from its nesting area. The killdeer is the only shorebird found year-round in most of the state. Its tracks—usually found in deep sand—are typical as well of the tracks of shorebirds that visit the region seasonally: the front center toe is longer than the two outer toes; the small rear toe, more of a heel spur than a toe, leaves a small imprint; and the tracks are usually printed in a line, only 1 or 2 inches apart for birds the size of the killdeer, less for sandpipers, and up to 6 inches apart for birds the size of a greater yellowlegs. The general shape and lack of any evidence of webbing between the toes separate shorebird tracks from those of gulls or ducks.

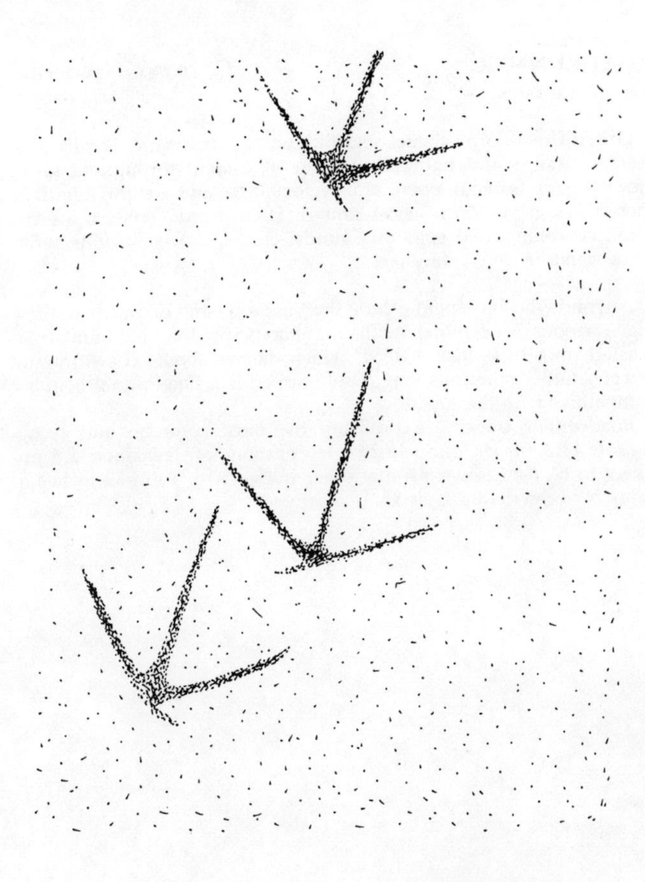

Killdeer
life size in sand

ROADRUNNER
Chaparral cock

Geococcyx californianus

Order: Cuculiformes (cuckoos, anis, and roadrunners). **Family:** Cuculidae. **Range and habitat:** year-round resident throughout lower elevations of Texas; in open, arid regions. **Size and weight:** length 22 inches, wingspan 27 inches; 1 pound. **Diet:** lizards, snakes, insects, mice, scorpions, and spiders. **Sounds:** clucks, crows, various series of descending notes, dovelike.

Anyone who has spent a little time in Texas and/or watched a few old cartoons is familiar with the habits of this medium-sized, crested, mostly terrestrial bird, which dashes about so swiftly and purposefully, trailing its long, white-tipped tail. One cannot help but sympathize with the coyote.

Roadrunner tracks are probably the most common and recognizable bird tracks found in the dry portions of Texas and are unlikely to be confused with any other tracks, with 10 to 12 inches or more between distinctively shaped prints.

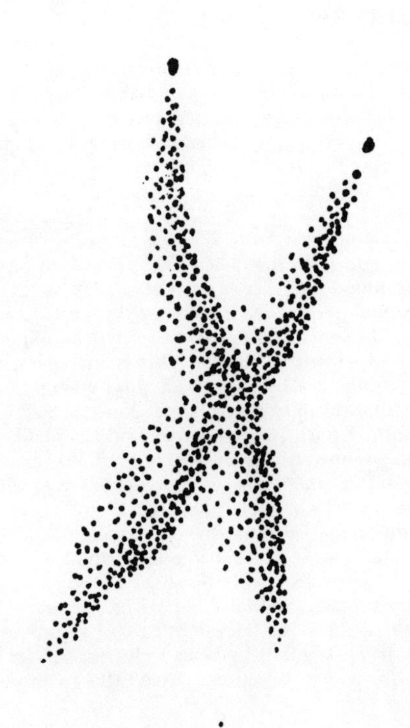

Roadrunner
life size in sand

TURKEY VULTURE *Cathartes aura*

Order: Falconiformes (vultures, eagles, hawks, and falcons). **Family:** Cathartidae (vultures). **Range and habitat:** year-round resident throughout Texas; commonly seen scavenging in fields and along roadsides nearly everywhere. **Size and weight:** length 30 inches, wingspan to 6 feet; 3 pounds. **Diet:** carrion. **Sounds:** grunts or hisses occasionally when eating.

The turkey vulture is a large black bird easily recognized by its naked red head and neck (or naked gray face of an immature bird), and by the two-toned wings that form a broad V as the bird soars in wide circles, looking for dead animals. Vultures usually soar in groups of several or many birds, and they will be joined by more as soon as they find something to eat. This is the characteristic which makes turkey vulture tracks easy to identify: a single track could be confused with that of an eagle or heron, but you will generally find hundreds of vulture tracks wherever you find any at all.

The track shape, however, with strong, thick toes and long, curved talons leaving significant prints, is representative of all the falcons, hawks, and bald and golden eagles that might be encountered in Texas, but none of the Falconiformes congregate in large groups on the ground; in fact, most rarely come to earth at all. Track size will vary according to species, with those of the eagles being slightly larger, while most hawks and falcons have feet about half this size. In the eastern half of the state, large numbers of slightly smaller tracks like these may have been left by black vultures, similar but more aggressive vultures, which sometimes drive turkey vultures away from a food source.

Turkey Vulture
life size in mud

GREAT BLUE HERON *Ardea herodias*

Order: Ciconiiformes (herons and allies). **Family:** Ardeidae (herons, egrets, and bitterns). **Range and habitat:** year-round resident of Texas; in most lowland areas, common on fresh-water and ocean shores. **Size and weight:** length 48 inches, wingspan 72 inches; 7 pounds. **Diet:** fish, snakes, insects, mice, and frogs. **Sounds:** "kraak" and strident honks.

The presence of a great blue heron magically transforms an aquatic landscape, adding an aura of quiet elegance characteristic of the best Oriental brush paintings. This large heron typically walks slowly through shallows or stands with head hunched on shoulders, looking for the fish that make up a large part of its diet. A heron's nest, maintained year after year, is an elaborate structure of sticks 3 feet across built in a tree; great blue herons often nest colonially.

You will most often find great blue heron tracks bordering the fresh-water areas where the bird feeds. The four toes and claws of each foot usually leave visible imprints. The well-developed rear toe enables the heron to stand for long periods of time on one leg or to walk very slowly while hunting.

This track shape is typical of all the herons, egrets, bitterns, and ibis found around Texas shores at various times of year. The size and location of the tracks will vary according to species, but the offset front toe pattern distinguishes this group of wading birds from sandhill cranes, rails, and other shorebirds of the state.

Great Blue Heron
life size in mud

Recommended Reading

CARE OF THE WILD FEATHERED AND FURRED: A Guide to Wildlife Handling and Care, Mae Hickman and Maxine Guy (Unity Press, 1973); unique perspectives on animal behavior and emergency care of injured and orphaned wildlife.

A FIELD GUIDE TO ANIMAL TRACKS, second edition, Olaus J. Murie (Houghton Mifflin Co., 1975); a classic work on track identification by Murie (1889–1963), an eminent naturalist and wildlife artist; one of the Peterson Field Guide Series; an excellent research text for home study.

A FIELD GUIDE TO THE MAMMALS, third edition, W. H. Burt and R. P. Grossenheider (Houghton Mifflin Co., 1976); a good general field guide.

ISLAND SOJOURN, Elizabeth Arthur (Harper & Row, 1980); an account of life on an island in British Columbia's wilderness, with a chapter devoted to a metaphysical perspective of animal tracks.

SNOW TRACKS, Jean George (E. P. Dutton, 1958); an introduction to the study of animal tracks for very young children.

THE TRACKER, Tom Brown and William J. Watkins (Berkley Publications, 1984); an intriguing story by a man who has devoted his life to the science of following tracks and other movement clues of various animals, including humans.

Index

About the author:

Chris Stall first became interested in wild country and wild animals as a Boy Scout in rural New York State, where he spent his youth. In the two decades since, he has traveled and lived around most of North America, studying, photographing, sketching, and writing about wild animals in their natural habits. His photos and articles have appeared in a number of outdoor and nature magazines. Stall is headquartered in Cincinnati, when not enjoying his favorite pastime, saltwater sailing.